The Farm Financial Crisis

The Farm Financial Crisis

Socioeconomic Dimensions and Implications for Producers and Rural Areas

edited by
Steve H. Murdock
and F. Larry Leistritz

88-414

Westview Press / Boulder and London

Westview Special Studies in Agriculture Science and Policy

Copyright © 1988 by Westview Press, Inc.

Published in 1988 in the United States of America by Westview Press, Inc.; Frederick A. Praeger, Publisher; 5500 Central Avenue, Boulder, Colorado 80301

Library of Congress Cataloging-in-Publication Data
The Farm financial crisis.
    (Westview special studies in agriculture science and
policy)
    Includes index.
    1. Agriculture—Economic aspects—United States.
2. Agriculture—United States—Finance. 3. Agricultural
credit—United States. 4. Agriculture and state—United
States. 5. Family farms—United States. I. Murdock,
Steven H. II. Leistritz, F. Larry. III. Series.
HD1761.F295 1988   338.1′3′0973   86-32601
ISBN 0-8133-7186-4

Printed and bound in the United States of America

The paper used in this publication meets the requirements of the American National
Standard for Permanence of Paper for Printed Library Materials Z39.48-1984.

6   5   4   3   2   1

Contents

Tables and Figures

Figures

Preface

After nearly a decade of prosperity, rural America entered the 1980s with its agricultural base facing a severe financial crisis. Land values, export markets and the general demand for agricultural commodities were declining while the levels of indebtedness reached during the 1970s were becoming increasingly difficult to manage. By the middle of the 1980s, the existence of a crisis was apparent in farm failure rates that had reached levels that had not occurred since the 1930s and in the fact that large numbers of agricultural banks were failing and agencies that provide loans to farmers and ranchers were experiencing unprecedented losses. Small towns in agriculturally dependent rural areas were losing businesses, populations and related services, and extremely high rates of socioemotional problems were noted among rural residents in agriculturally dependent areas of the nation.

Although it was clear that a farm financial crisis was gripping rural America, it was less clear how pervasive the crisis was, what the characteristics were of those most likely to experience the direct and indirect impacts of the crisis, what secondary impacts were occurring in the rural communities dependent on agriculture and what the long-term consequences of the crisis were for agriculturally dependent rural areas. The literature available to address such issues was growing rapidly, but it was characterized by national-level analyses that concentrated only on the farm sector, by case-study analyses of individual farming and ranching areas or communities in single states, by some data on producers but little more than conjecture about the secondary impacts on rural communities, and by an emphasis on describing present trends rather than evaluating both short-term and long-term consequences of the crisis for agriculturally dependent rural areas in the United States. Comprehensive analyses of the socioeconomic impacts of the crisis were simply not available.

This work is an attempt to address the need for more comprehensive analyses of the socioeconomic impacts of the farm financial crisis of the 1980s. The authors examine the socioeconomic context, the impacts and the long-term consequences of the crisis, emphasizizing how these affect agriculturally dependent rural areas. In addition to the survey of the base of empirical data available by the late 1980s, the results of the editors' own extensive analyses of the impacts of the crisis on the producers, former producers, business operators, former business operators, employees of rural businesses, and other rural residents are used. While this analysis is sometimes limited by data specific to given regions of the nation, the intent is to develop generalizations useful for discerning the impacts of the crisis in areas throughout the United States, for delineating policy options that address the short-term consequences of the crisis, and for identifying research and long-term policies needed to create the information base and resource base essential for resolving the long-term problems.

The work is intended for use by researchers, policy makers, and others interested in the farm financial crisis of the 1980s and in the long-term social and economic development of rural America. That the work is restricted by the very limited base of available data on the crisis and by the even more limited base of knowledge on the inter-relationships between changes in the farm sector and those in rural communities is evident. Our hope, however, is that by delineating what we have learned about the characteristics of the crisis, and about the implications of a declining agricultural base in general, and by delineating the research needed to address what we do not know, we can assist in the development of policies that will reduce the probabilities of future crises and help preserve the agricultural, economic, demographic and sociocultural base of agriculturally dependent areas in the United States.

Steve H. Murdock
F. Larry Leistritz

Acknowledgments

In the completion of this work, the support, encouragement and assistance of numerous persons and agencies have been invaluable. The Department of Rural Sociology and the Texas Agricultural Experiment Station in the Texas A&M University System, the Department of Agricultural Economics at North Dakota State University, the North Dakota Agricultural Experiment Station, the North Dakota Cooperative Extension Service, and the Otto Bremer Foundation all provided financial support for this effort and receive our sincere appreciation. We also extend appreciation to the Agricultural and Rural Economics Division of the Economic Research Service of the U.S. Department of Agriculture for providing financial assistance for part of the research reported in this volume and especially to Dr. Fred Hines from that agency, who provided constructive criticism and useful suggestions that improved the quality of that research.

In the preparation of the book, numerous persons have provided assistance in data collection and analysis, in manuscript preparation and in providing critical reviews of the volume. In this regard, coauthors Don E. Albrecht, Kenneth Backman, Brenda L. Ekstrom, Rita R. Hamm, Lloyd B. Potter, Harvey G. Vreugdenhil and Janet Wanzek deserve our sincere appreciation. We also wish to thank Arlen G. Leholm, Gary Goreham, Richard Rathge, Tim Mortensen and Eldon Schriner for providing helpful comments on the design of the research reported in this volume. In like manner, the extensive programming efforts of Darrell Fannin and Harvey G. Vreugdenhil are gratefully acknowledged. Special appreciation is extended to Donna Nunez, who showed a unique level of dedication and patience in typing numerous drafts of the volume. We also wish to thank Sue Bartuska, Tonia Roberts, James DeGuzman, Debbie Tanner and Phyllis Owens, who assisted in typing and preparing figures in several drafts of selected chapters, and Patricia Bramwell, who displayed her usual high level of dedication in supervising staff in the preparation of the manuscript. Our appreciation is also extended to Beverly Pecotte for her persistence in checking and rechecking references in the text as frequent revisions were made. We express special appreciation to Brenda L. Ekstrom, who provided extensive editing suggestions for each page of the manuscript that substantially improved its quality.

Our highest level of appreciation, however, is extended to Rita R. Hamm, who served as the lead editor for this volume. Her dedication and long hours of effort in not only ensuring that the text was grammatically correct and properly punctuated but in ensuring that each page was properly transcribed for computer typesetting and each figure properly placed for reproduction, as well as in providing useful suggestions for substantive changes, were instrumental to the completion of the project. We owe her a special debt of gratitude.

Finally, to our colleagues, staff and friends who patiently endured our impatience and our neglect of joint projects during the time we dedicated ourselves to the completion of this volume, we extend our thanks for their tolerance and understanding.

S.H.M.
F.L.L.

Introduction

Steve H. Murdock and F. Larry Leistritz

American farmers are facing their most severe financial crisis since the 1930s. As a result of low commodity prices, high interest rates, and falling land values, a large number of farmers have left farming and many more are likely to be forced to quit farming in the next 2 to 5 years. The U.S. Department of Agriculture (Johnson et al. 1986) estimates that 21 percent of all farmers have debt-to-asset ratios of over 40 percent, a ratio considered to be high and to cause severe financial stress. Other analyses (Albrecht et al. 1987a; Murdock et al. 1985; Leistritz et al. 1985a; Richardson and Bailey 1982a, 1982b) suggest that nearly a third of farm operators in the agriculturally dependent parts of the Midwest and Plains states (i.e., counties with 20 percent or more of their income and employment in agriculture; see Bender et al. 1985) may have such debt-to-asset ratios. In addition, data from recent analyses (Federal Deposit Insurance Corporation 1987; University of Colorado 1985; Meekhof 1983) indicate that farm bankruptcies and foreclosures are occurring at a rate that is several times the historical average. Equally important, available information (Murdock et al. 1986a; Leistritz et al. 1985a; Richardson and Bailey 1982a, 1982b) suggests that the farmers being forced from their farms are those in the most productive ages who are operating family farms; that is, farms of moderate size whose operators earn a majority of their income from farming (Bultena et al. 1986; Murdock et al. 1986a; Heffernan and Heffernan 1985b).

The failure of a large number of farm enterprises has many implications for the structure of agriculture in the United States. It will likely lead to a further increase in the dualistic nature of agriculture in the United States (Albrecht and Murdock in press; Larson 1981), resulting in a further reduction in medium-size farm operations. It may alter ownership and tenant patterns along with the historic patterns of increasing farm consolidation. It is unlikely, however, to substantially affect the overall

1

level of production of food and fiber in the United States or to affect
consumer prices for food products. Thus, the effects of farm failures on
national trends in agriculture will be complex and slow to evolve, but
the impacts of such farm failures on farm operators and rural communities
can be expected to be evident immediately.

Farm failures affect farm operators, their families, and the communities
and businesses dependent on farm populations. Many farm operators
are undergoing extensive social and psychological stress and, if displaced,
farm operators will be forced to seek new forms of employment, often
in communities located far from their present residence (Albrecht et al.
1987b; Bultena et al. 1986). Their families are experiencing increased
stress during the failure of the farm enterprise and will face adjustment
problems, if they must relocate (Leistritz et al. 1987a).

Many rural, agriculturally dependent communities are experiencing
numerous problems because a significant proportion of the farm families
in their area are being forced to leave (Murdock et al. 1987b). Their
financial institutions and other businesses have had reduced business
volume, and some businesses have been forced to close (Murdock et
al. 1987b). This, in turn, has resulted in the relocation of a substantial
proportion of nonfarm residents from rural communities to other locations
(Leistritz et al. 1987a). In like manner, many rural services face declines
in service populations of a sufficient magnitude to threaten their continued
economic viability. For those rural communities in the United States
that are most dependent on agriculture, the present rate of farm failure
may be sufficient to threaten their viability as trade and service centers.

Finally, the present situation threatens the existence and/or viability
of many farm-oriented financial corporations and institutions, both
private and public. The proportion of loans that cannot be repaid has
led to the failure of numerous agricultural banks (Federal Deposit
Insurance Corporation 1987) and has reached a level in others that is
likely to threaten the financial stability of these institutions. The failure
of such institutions results in additional instability in the rural economy.

In sum, the present rate of farm failure will have substantial economic,
demographic, and social impacts on farm operators, on rural communities,
and on community residents in agriculturally dependent communities.
If policies are to be designed to address this crisis and to assist farm
operators and communities in adjusting to these problems, information
on such impacts and adjustments is essential. Although recent recognition
of the problematic nature of the current farm financial situation has
resulted in numerous articles and reports (Doeksen 1987; Murdock et
al. 1987a, 1986a; Bultena et al. 1986; Office of Technology Assessment
1986; U.S. Department of Agriculture 1985), these studies have generally
concentrated on estimating the percentage of farm operators who are

experiencing financial stress or on examining a limited number of impacts at single sites (Heffernan and Heffernan 1985a; Murdock et al. 1985). In like manner, although numerous general works detailing and criticizing the structure of American agriculture exist (Busch and Lacy 1983; Larson 1981; Breimyer 1977), few have presented detailed information on either the community-level impacts of farm failure or on the processes involved in the transition from farm to nonfarm employment (Doeksen 1987; Murdock et al. 1987b). The determinants and the impacts of the current farm financial crisis and of farm failure in general have thus received limited attention in the literature to date. No comprehensive overview addressing the determinants, concomitants, and secondary socioeconomic impacts of the current farm financial situation has been completed.

In fact, although high levels of farm failure and rural population decline have occurred since the 1930s and the number of farms and the extent of population loss are well established (Brown and Beale 1981), a recent analysis of one of the major effects of farm decline— population loss in agriculturally dependent communities—noted that:

> Agricultural service towns and villages are by far the most numerous case of declining settlements in the United States, for the past century or so and for the present . . . The frequency—indeed, the ubiquity—of population declines in the United States is not well appreciated, even by demographers, geographers, economists, and other students of population change, land occupancy, and resource use. In our obsession with growth, we have ignored or underestimated the declines. In my judgment, no really comprehensive and adequate analysis of population decline in the United States has yet been made (Clawson 1980: 67, 69).

Thus, relatively little historical or recent information is available on either the determinants or the socioeconomic impacts of farm failure.

The lack of comprehensive examinations of the determinants and consequences of the processes underlying and resulting from farm failure are thus a product of a long-term pattern of neglect. For the recent crisis, however, evidence related to the apparent impacts of the crisis has begun to accumulate in just the last few years. Only recently has evidence appeared that makes a more comprehensive examination possible. At the same time, it is clear that additional factors have also led to a relative neglect of the topic in the professional literature.

A second reason for the lack of information on the determinants and consequences of the crisis lies in the fact that many of the conditioning factors that have led to the present crisis have been slow to evolve, and thus changes in these factors have been easily ignored or seen as merely a continuation of historical patterns. For example, the present crisis can generally be traced to the economic trends and policies of the 1970s and early 1980s, but the structure of U.S. agriculture, which is being impacted by the crisis, and which is, in turn, accentuating the impacts of the crisis, has evolved over at least the past fifty years. In like manner, the changes that have taken place in the financial structure of agriculture and in rural populations have occurred over several decades. Yet, these factors have affected many of the characteristics of the present crisis and have been affected, in turn, by the crisis. In sum, many determinants and factors conditioning the crisis have simply not been examined because they have been seen as merely an accentuation of long-term trends.

A third reason for the relative absence of examinations of the farm crisis lies in the lack of good empirical data on the consequences of farm failure. Although numerous examinations of the general magnitude of the present crisis in different regions of the United States are appearing (Doeksen 1987; Bultena et al. 1986; Leistritz et al. 1986b, 1985a; Leholm et al. 1985a; Murdock et al. 1985), there are relatively few comprehensive examinations of the consequences of high rates of farm failure that attempt to formulate a multiregional picture of the crisis. Rather, most analyses tend to be examinations of only a few social factors (Heffernan and Heffernan 1985a, 1985b) or of farm structure (Murdock et al. 1986a) or to be estimates based on assumptions obtained from historical patterns (Doeksen 1987). Among the most obvious needs related to understanding the present crisis is the need for good empirical data. In particular, detailed information is needed on which types of producers are being most affected by the crisis and on the ways producers are responding to the crisis (in terms of production procedures) and adapting to its financial, economic, and social consequences. Information on the economic, social, demographic, and farm structure characteristics and on the problems producers are experiencing has simply not been examined adequately in works available to date. In addition, data are not available on the short-term and long-term impacts of farm failure on the demographic, economic, and social structure of rural areas. In sum, works that provide an improved data base are needed for understanding the present and long-term determinants and consequences of the farm crisis.

A fourth reason for the relative neglect of the crisis lies in the fact that the impacts of the crisis on rural areas are being compounded by a myriad of other factors that are also having profound effects on rural economies. In many parts of the country, for example, federal cutbacks in funding for services, the decline of energy-producing industries, and a generally depressed financial situation are leading to substantial changes. It is extremely difficult to differentiate those changes due to other factors from those due to problems in the farm sector. It is very important to attempt to systematically trace the potential effects of the present crisis on the economic, demographic, and social structures of rural areas within the context of other factors affecting rural areas.

In sum, literature examining the determinants, concomitants, and consequences of the present farm crisis is limited due to a variety of factors. There is a critical need for works that provide an examination of the historical and current conditions in rural areas that form the context for the present crisis, for works that examine the characteristics of those who are being most affected by the present crisis, and for works that examine the crisis' short- and long-term impacts on producers, production units, and on the economic, demographic, and social characteristics of rural communities.

The purpose of this book is to address these needs. Specifically, three broad aspects of the present crisis are examined: (1) the historical and current characteristics of U.S. agriculture and of rural areas in the United States that have formed the context for the development of the present crisis; (2) the financial, demographic, economic, and social characteristics of farm operators experiencing the most substantial levels of farm financial stress; and (3) the current and long-term impacts of the crisis on producers and production units and on the economic, demographic, social, and community service characteristics of rural areas. In addressing the first of these three aspects, the work draws heavily on the historical literature on the structure of agriculture, on rural population change, and on the financial structure of agriculture. The examination of the second and third issues draws from existing data and analyses of the farm crisis. In this discussion extensive use is made of our analyses of producer and rural community populations in the states of Texas and North Dakota (Leistritz et al. 1987a, 1986a; Murdock et al. 1987a, 1986a, 1985a; Leholm et al. 1985a). In the final chapters of the work, we also develop a set of assumptions about the severity and continuity of the current crisis and about its likely impacts on various dimensions of rural areas and then trace the long-term implications of these assumptions for rural areas. The work thus provides one of the first attempts to examine the determinants and the current and long-term consequences of the crisis for producers and for rural areas in the United States.

Organization and Content of the Work

The work is organized into this introduction, a concluding section on policy alternatives/implications, and seven substantive chapters organized into two parts. The chapters in the first part describe the context and determinants of the crisis. This section contains three chapters, each of which examines the general historical context of a specific set of factors directly related to the farm crisis. Chapter 1 examines the financial characteristics of farms in the United States, specifically the financial and economic patterns and policies that engendered the farm crisis. Chapter 2 examines the structure of agriculture by describing the historical patterns of change in such factors as farm size, farm tenancy, and technology use and how such characteristics have conditioned distinct response patterns to the present crisis. The third chapter presents an overview of the demographic, socioeconomic, and public service conditions in rural areas. Together, the chapters in Part 1 attempt to provide a background for understanding the determinants of, and the factors affecting and being affected by, the current farm financial crisis.

The second part of the work examines the characteristics of the crisis itself and its impacts on farm producers and rural populations. It contains four chapters intended to describe the types of production units (farms and ranches), the characteristics of those producers experiencing different levels of financial stress, and the reactions and adaptations of producers and rural areas to the crisis. The first chapter in this section, chapter 4, examines findings from analyses that describe the financial conditions of farm producers in various parts of the country. This chapter examines the debts, assets, and other characteristics of farms in various financial circumstances. Its intent is to provide definitive evidence of the actual levels of debt and characteristics of production units under financial stress in different parts of the United States and to establish the characteristics of such units.

The next chapter in this part, chapter 5, examines the characteristics of producers undergoing various levels of financial stress and describes the means by which producers are reacting and adapting to financial stress. It examines current production changes resulting from attempts to adapt to the crisis and the potential implications of such changes for producers and production units. This chapter uses data from analyses from across the country as well as the results of detailed analyses of data for sites such as Michigan, Missouri, Texas, and North Dakota to assess a wide range of producers' reactions and adaptations. It includes an analysis of financial and production changes being made by producers in response to financial pressures.

The age, levels of education, family characteristics, employment histories, and other characteristics of producers experiencing various levels of stress are examined in chapter 6. This chapter also provides an examination of off-farm employment and other means being used to establish alternative sources of income, and presents an evaluation of the social and psychological reactions of producers to the crisis. Finally, chapter 6 examines the social impacts of the crisis on producers and other residents in rural areas and the impacts on rural communities.

The final chapter, chapter 7, examines the implications of the crisis for the structure of agriculture in the United States and the economic, demographic, community service, fiscal, and social implications of the crisis for rural America. Using a set of assumptions about rates of farm failure and general rural population loss due to the crisis, this chapter attempts to provide information concerning both the types and magnitudes of immediate and long-term implications likely to be experienced by agriculturally dependent rural areas in the United States. This chapter evaluates the potential effects of alternative levels of farm failure on the structure of agriculture; on the level of business activity; on rural populations; on the continued viability of agriculturally dependent communities; on the quantity, quality, and distribution of services in rural areas; and on the fiscal base to support community services. Overall, the four chapters in Part 2 attempt to establish the magnitude and characteristics of the crisis, the characteristics of producers and the families of producers experiencing various levels of financial stress, and the immediate and potential long-term reactions and adaptations to the crisis by both producers and other rural residents.

The *Policy Alternatives* section summarizes the key points of the work and discusses the policy implications of the crisis. Special emphasis is given in this chapter to delineating policy alternatives that can ameliorate the immediate and long-term consequences of the crisis for rural areas. Finally, this chapter outlines the types of policy-related data and research necessary to further establish the determinants and consequences of farm financial changes.

Overall, then, the work is organized to provide the reader with an overview of the determinants, the concomitants, and the immediate and long-term impacts and implications of the farm crisis in the United States. It is intended for use by academics, policy makers, and others interested in the development of agriculture and rural areas in the United States.

Limitations of the Work

The work is obviously limited in several regards. First, it focuses only on socioeconomic dimensions. In fact, the work examines only those dimensions as they relate to producers, production units, and rural areas. It makes no attempt to examine the implications of the crisis for the overall production of food and fiber in the United States, for national or international commodities markets, or for national financial markets or institutions. Its focus is admittedly, then, localistic and selective.

A second limitation lies in the focusing of the analysis on the current farm financial crisis because of the ambiguity surrounding both the existence and the definition of the farm crisis. The term "financial crisis" clearly suggests that an unusually large proportion of producers have excessive levels of debt and inadequate farm incomes, but there is no uniform agreement among scientists or policy makers as to the financial situation or the proportion of producers who must be in such a situation for a crisis to exist. Although we, like other social scientists, often use the existence of a debt-to-asset ratio of 40 percent or more as indicative of financial management difficulties and believe debt-to-asset ratios over 70 percent are likely to lead to farm failure, it must be recognized that such categories are arbitrary. Thus, we rely on the fact that levels of farm failure and indebtedness are high, and that there is relatively widespread consensus that such are at crisis proportions to legitimize our use of the phrase "farm financial crisis." Equally important, however, is the fact that the current crisis is accentuating long-term economic problems in rural areas. Tracing the interrelationships between such problems and the current crisis is vital for understanding the future of rural America. Our interest in the crisis is not in a phenomena that may be evident only in a few years of accentuated rates of farm failure but in its long-term impacts on the future of rural America. An examination of the current crisis is essential for discerning not only the current but also the future socioeconomic characteristics of rural areas in the United States.

The work is also limited to an examination of the farm financial crisis in the United States. Even a cursory examination of international agriculture (see Wilkening and Galeski in press) suggests that agricultural producers in many other nations of the world are experiencing conditions similar to those faced by producers in the United States. In examining the crisis only in the United States, we do not intend to imply that such crises have occurred only in the United States or that they have been more severe in the United States. Rather, space and other limitations, as well as the interests of the authors, simply resulted in our limiting the analysis to the United States.

A fourth and perhaps the most obvious limitation of the work is simply that neither the historical nor the current data available are entirely adequate for a detailed examination of the topic of the work. To perform the analysis, we use data from scattered sources and from our own work. Throughout, we have been forced to speculate more often than we would like and to draw implications from what are admittedly very limited data bases and analyses. Although these limitations are severe, we believe that the limitations in the data available to address a topic are often best demonstrated by attempting to draw existing fragmented data sources together to display the present state of knowledge or ignorance related to a topic. It is our hope that the work not only highlights the limitations of available data on the crisis, but also clearly delineates the data required to adequately analyze the socioeconomic aspects of the crisis and its long-term impacts on rural America.

Finally, the work reflects the biases of the authors. These biases lead us to perceive the increasing rate of farm failures and related dimensions in rural areas as largely negative. That such changes may, in fact, lead to a more efficient agricultural base or even perhaps to a greater level of overall production is not a subject considered in this volume. The declining agricultural base in rural areas is, to these authors at least, a subject of concern and a trend that seems unlikely to lead to a more desirable rural America for either its present residents or American society.

Overall, then, this work is limited in many regards. It reflects the problems, and we hope the prospects, of one of the first works on a topic for which both the definition of the problem and its associated data are poorly developed. By focusing on the socioeconomic dimensions of the farm crisis and its short- and long-term implications for producers and rural areas, we hope to stimulate additional analyses. Even more important, we hope to assist readers, in general, in gaining a better understanding of the determinants and consequences of the crisis and to assist policy makers, in particular, in gaining the insights necessary to develop policies to address the current crisis and the long-term problems in rural America.

PART ONE

The Context of the Crisis

1

Financial Characteristics of Farms and of Farm Financial Markets and Policies in the United States

F. Larry Leistritz and Steve H. Murdock

In less than a decade, many areas of rural America have undergone a drastic reversal of fortunes. From a period of unparalleled growth in asset values and farm income and widespread optimism concerning the future, the farm sector has been plunged into what is, in many respects, its most severe economic stress since the 1930s. Land values have fallen precipitously—in several major farm states by more than 50 percent since 1981—and net farm income (adjusted for inflation) has receded to only a fraction of the levels attained in the mid-1970s. The frequency of farm foreclosures, forfeitures on land contracts, and defaults on notes has reached levels not seen since the days of the Great Depression (Harl 1986). Ever-increasing numbers of farm families have been forced out of business and face the prospect of leaving their home communities to seek work in a distant city (Leistritz et al. 1987a).

Further, economic stress is not confined to indebted farm families but now affects almost all segments of many agriculturally dependent rural communities. As farm families have been forced to defer purchases, local businesses have suffered declining sales, and many have been forced to lay off workers and to retrench in other ways (Leistritz et al. 1987b). The result has been failing businesses, declining employment, unpaid property taxes, and reduced ability to support government services in many rural areas (Stinson et al. 1986).

The series of economic changes leading to this reversal of fortunes, as well as the causal forces underlying them, is the subject of this chapter. First, the economic and financial changes that have driven many farm families to, and in many cases over, the brink of bankruptcy are examined, and the current financial status of the farm economy is briefly summarized. The forces underlying these changes are then analyzed.

The Changing Economic Environment of American Agriculture

Farming, ever a dynamic and inherently risky economic endeavor, became even more so over the past two decades as a combination of forces combined to put the U.S. agricultural sector on an economic roller coaster. Although identifying a single cause of current financial conditions would run the risk of oversimplification, much of the current distress among farmers and agricultural lenders has its roots in excesses induced by the inflationary conditions of the 1970s coupled with optimistic expectations concerning the worldwide demand for farm products.

The decade of the 1970s was one of rapid growth in exports of U.S. agricultural commodities, which increased at more than 8 percent per year (Harrington and Carlin 1987). The growth in exports was stimulated by a favorable climate of economic growth worldwide, and particularly in many developing countries, and by the declining value of the U.S. dollar, which made U.S. products progressively cheaper to foreign buyers. Farm prices rose substantially in the early 1970s in response to a strong export demand for wheat and feed grains. Both market signals and federal government policy encouraged farmers and agribusinesses to expand production, and some policymakers foresaw the disappearance of excess capacity in agriculture (for example, see Lee 1982).

Many farmers expanded their operations by investing heavily in new machinery, equipment, and, most significantly, land. Such investments were favored not only by increased farm profits but also by very low and sometimes negative real interest rates (the real interest rate is the nominal interest rate minus the inflation rate). Farmers borrowed heavily to finance these investments. Farm debt rose more than 10 percent per year and had tripled by 1980 (Figure 1.1). Land values increased much more rapidly, however, partly in response to pressures for farm expansion and partly because land was seen as a good hedge against inflation (Barrows et al. 1986). As land values and farmers' equity (net worth) positions increased, farm lenders often encouraged additional borrowing to finance expansion. Through this period of rapid expansion, U.S. agricultural production surged, and agribusinesses and farm-based communities generally prospered.

By the early 1980s, the forces that had driven the rapid economic expansion of agriculture had reversed direction. A worldwide recession and a sharp rise in the value of the dollar led to a reduction in the export demand for U.S. farm products. At the same time, relatively high

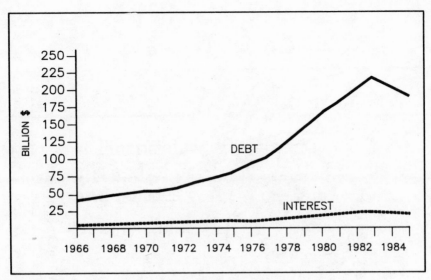

SOURCE: USDA 1986, Economic Indicators of the Farm Sector

Figure 1.1   Total farm debt and interest payments, 1966–1985.

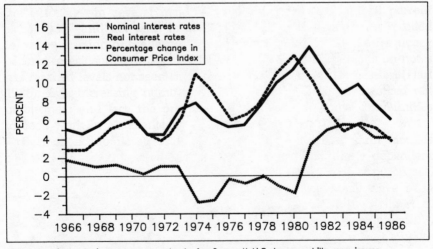

Note: Nominal rate is average annual rate for 6–month U.S. treasury bills, new issues.
        Real rate is nominal rate minus yearly percentage changes in the Consumer Price Index.
SOURCE: Economic Report of the President 1987.

Figure 1.2   Nominal and real interest rates and yearly percentage changes
            in the Consumer Price Index.

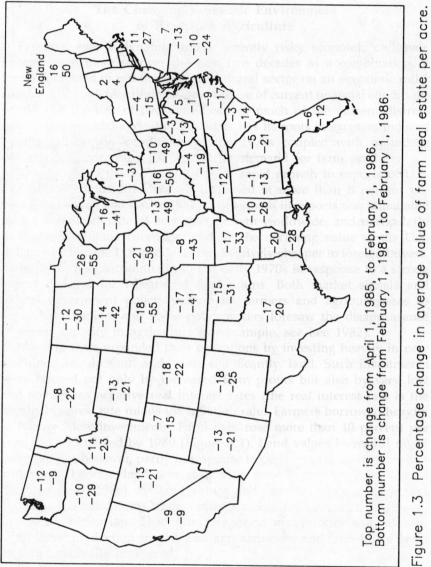

Figure 1.3 Percentage change in average value of farm real estate per acre.

Top number is change from April 1, 1985, to February 1, 1986.
Bottom number is change from February 1, 1981, to February 1, 1986.

loan rates for key U.S. farm commodities provided incentives to other countries to substantially increase their production capacity, and former foreign customers entered the world market as U.S. competitors. By 1985, U.S. farm exports had fallen by 33 percent from their peak in 1981, and prices of most major commodities had dropped sharply (Petrulis et al. 1987).

At the same time, the costs of producing farm commodities increased. As tight monetary controls were applied to stifle inflation, real interest rates climbed to unprecedented levels in the early 1980s (Figure 1.2). Costs of other farm inputs, such as fuel and fertilizer, were generally increasing during this period, and the real value of net farm income, while fluctuating sharply, showed a general downward trend. As declining exports and commodity prices and rising costs reduced the returns to land and created the expectation that future returns would be even lower, farmland values began a major decline. By 1986, farmland values for the nation as a whole had declined 27 percent from their 1981 peak, and in several major agricultural states values had fallen almost 60 percent (Figure 1.3).

As farmland values declined, farmers' equity positions eroded in general. For some, asset values dropped below the total of their liabilities, and others found that their diminished farm income was no longer adequate to service (i.e., make interest and principal payments) the massive debts accumulated during the boom period. As lenders refused further credit to some borrowers who appeared to be insolvent or initiated legal action against those who were delinquent on loan payments, farm liquidations increased. From 1982 to 1986, the percentage of farms going out of business nearly tripled (Table 1.1), and the percentage going through bankruptcy more than quadrupled.

The period since 1981 thus has been one in which the balance sheet of U.S. agriculture has undergone substantial changes. The value of total farm assets has decreased by almost one-fourth (Table 1.2), and almost all of this decrease reflects reductions in real estate values. Levels of farm debt have changed little during this period, and so the reduction in asset values has been directly reflected in equity levels. However, the sector's debt-to-asset ratio of .25 in 1985 was still substantially lower than that of most other industries, so from that perspective it can be argued that its overall financial position is reasonably sound.

The sector's average figures, however, conceal considerable diversity among farmers and ranchers. As of the end of 1985, about 78 percent of farmers nationally had debt-to-asset ratios of less than .40. Generally, it has been thought that most of these farmers will be able to service their debt and pay other costs when due even under the economic conditions prevailing in the mid-1980s. However, the remaining 22

Table 1.1

Farms Going Out of Business or Going Through Bankruptcy, 1982–1986

Number/Reason	1982	1983	1984	1985	1986
	(– – – – – – percent – – – – – –)				
Farms going out of business	2.2	2.3	3.6	4.8	6.2
Farms going through bankruptcy	.8	1.1	2.6	3.8	4.2
Why farmers are going out of business:					
Normal attrition	N/A	37.7	31.3	27.7	28.9
Voluntary liquidation	N/A	42.4	44.0	44.3	41.7
Legal foreclosure	N/A	18.1	22.3	25.8	26.3
Other	N/A	1.8	2.4	2.2	3.1

N/A = Not available.
Note: Years end June 30 of the year specified.

Source: USDA 1986a.

Table 1.2

Summary of the Farm Sector Balance Sheet, December 31, 1981 and 1985

Balance Sheet Items	1981	1985	Percentage Change
	(– – billion dollars – –)		
Assets:			
Total farm assets	1,005.2	771.4	−23.3
Real estate[1]	780.2	559.6	−28.3
Other	225.0	211.8	− 5.9
Total farm debt	188.8	192.1	1.7
Real estate[1]	97.2	97.3	0.1
Other	91.6	94.8	3.5
Equity	816.4	579.3	−29.0
Debt-to-asset ratio	.188	.249	32.4

[1]Excludes operators' dwellings.

Source: USDA 1986b.

percent of the farmers and ranchers (i.e., those with debt ratios over 0.40) owed about 66 percent of the total farm debt (Johnson et al. 1986). Somewhat more than half of these producers also had inadequate income from all sources in 1985 to cover current farm operating expenses, a family living allowance, and principal payments. This subgroup alone accounted for 37 percent of the total farm debt.

The substantial percentage of farm debt held by highly leveraged operators and those with inadequate cash flows indicates that the impact of debtor distress on lenders may be considerable (Harl 1986; Todd 1985; Melichar 1987). As indicated in Table 1.3, commercial banks, units of the Farm Credit System (i.e., Federal Land Banks and Production Credit Associations), and the Farmers Home Administration (FmHA) are the largest lenders to the farm sector. The impact of farm financial problems on some of these lenders already has been substantial. In 1985, the Farm Credit System sustained a $2.7 billion loss, the largest one-year loss of any U.S. financial institution, and 68 of the 120 banks that failed in 1985 were agricultural banks. The concentration of debt among the most heavily indebted farmers indicates that further deterioration of the financial condition of lenders is almost certain to occur (Harl 1986).

Causes of Farm Sector Problems

Underlying the traumatic adjustments that the farm and agribusiness sectors have been experiencing are a number of basic causal factors. Although these factors are intertwined, they can be viewed as a convergence of national and international economic trends, federal monetary and fiscal policies, and evolutionary changes in the structure of U.S. agriculture. These factors are discussed in the sections that follow.

National and International Economic Trends

A combination of long-term (or secular) and short-term (or cyclical) factors have contributed to the present state of the U.S. farm economy. One of the most fundamental of these forces is a continuing growth in agricultural productivity, not only in the United States but also around the world. Over time the result has been that the capacity to produce most agricultural products has grown more rapidly than the demand for them. This change has led to depressed farm prices and signaled a need for fewer resources in the agricultural sector (for additional discussion, see Baker [1987] or Heady [1962]). Because of recent advances in agricultural productivity, many less-developed nations now import fewer agricultural commodities. This trend may signal an increased

Table 1.3

Debt Owed by Farm Operators

Lender	Debt-to-Asset Ratio, January 1, 1986				
	.01–.40	.41–.70	.71–1.00	> 1.00	Total
	(– – – – – – – – million dollars – – – – – – – –)				
Commercial banks	12,007	10,508	4,284	4,263	31,072
Federal land banks	8,164	8,936	5,380	2,663	25,142
FmHA	2,626	4,833	3,538	6,035	17,082
Production credit assns.	3,704	2,951	1,116	1,037	8,807
Commodity credit corps.	2,652	2,988	1,467	1,146	8,253
Other individuals	5,042	3,950	2,092	1,544	12,628
Others	2,847	2,378	1,089	823	7,136
Merchants and dealers	766	446	317	330	1,860
Other farmers	386	258	410	364	1,419
All farms	38,195	37,248	19,692	18,205	113,389[a]

[a]USDA acknowledges that the figure given for "operator debt" is about $91 billion less than that for the sector with about $39 billion in "unexplained differences" (Johnson et al. 1986: 33).

Source: Johnson et al. 1986.

pressure for agricultural adjustments in countries such as the United States that traditionally export substantial amounts of farm products (Avery 1984).

The agricultural adjustment problem caused by growth in productivity is coupled with yet another problem—the tendency for the income elasticity of demand for food products as a group to decline as per capita incomes rise. This pattern occurs because, as nations experience economic development, their populations spend a smaller proportion of their income for food, and it suggests that, as economic development continues, a trend of depressed prices and earnings in the farm sector relative to other sectors of the economy can be expected.

These long-term trends do not spell unrelieved depression for the farm sector, however. Historically, the sector has experienced a number of short episodes of rising relative prices and improved returns to resources. Three of the most recent of these periods occurred in the decades beginning 1910, 1940, and 1970. Each period was characterized by a unique combination of forces that either stimulated the demand for farm products, disrupted traditional supplies, or both. As a result, farm prices increased, and returns to resources in agriculture improved substantially. However, a typical result of such favorable periods has been substantial commitments of additional resources to agricultural production and bidding up the price of farmland. Thus, relatively short periods of favorable returns appear to create expectations that these returns will be permanent. Then, as the short-term conditions that sparked the boom run their course, the farm sector finds itself with excess capacity and inflated land values, and an extensive period of depressed farm earnings generally follows. (For additional discussion, see McKinzie et al. [1987]).

Accentuating the effects of these long-term factors were a series of shorter-term forces that combined to first foster growing farm exports and higher prices during the 1970s and then depress the sector during the 1980s. First, the 1970s was a period of strong economic growth internationally. Increasing incomes led to a desire to improve diets, particularly in the developing countries of the world. From 1972 to 1979, the world trade in food and feed grains and in oilseed crops grew by an average of 8 percent per year, and the United States was able to capture a growing share of that market.

A major reason why the U.S. share of farm exports grew was that the value of the U.S. dollar was falling during the 1970s. The decline in the value of the dollar had the effect of making U.S. commodities less expensive to foreign customers, relative to those of many competing suppliers.

These conditions were reversed early in the 1980s, however, and the impact of these cyclical changes on the farm sector soon became apparent. Growing concern about accelerating inflation led the United States and other industrialized countries to adopt policies of economic restraint to restore price stability. The ensuing recession was the most severe since the 1930s. From 1979 to 1982 there was essentially no growth in real output in the United States, while output for the other major developed nations as a group fell by 8 percent. Interest rates rose to record levels, and many lesser developed countries experienced difficulty meeting their interest payments to foreign banks.

One result of these recessionary conditions was a dampening of the demand for farm commodities. From 1979 through 1984, the total volume of trade in feed and food grains and in oilseeds grew by only 1.6 percent annually, compared with an annual growth of 8 percent for the period 1972–1979. In addition, since 1981 the U.S. share of the total value of world exports has fallen, as has the value of U.S. exports (Figure 1.4). The decline in the U.S. share of world farm exports is attributable in large measure to a sharp rise in the value of the dollar, which in turn made U.S. commodities more expensive relative to those from other sources.

It can be contended, then, that major national and international economic trends have been among the primary causes of the current problems faced by American agriculture and that changes in these factors since the 1970s have resulted in large measure from changes in federal government policies.

Federal Government Policies

Changes in federal monetary and fiscal policies, in policies regulating financial institutions, and in policies directly affecting farm prices and incomes have all played a substantial role in determining the current economic and financial status of U.S. agriculture. It can be argued that the most significant single event influencing the current economic status of agriculture was the decision by the Federal Reserve Board in October 1979 to limit the supply of credit in order to curb inflation in the U.S. economy (Harl 1986). This action rapidly pushed interest rates to very high levels, and inflation dropped substantially. The effect of the tight money policy was compounded by the government fiscal policy embodied in the Economic Recovery Tax Act of 1981. The tax cuts included in that bill virtually assured massive federal budget deficits.

Because a budget deficit can only be financed by borrowing, the federal government was forced to enter the money market to borrow increasing amounts of capital, and further upward pressure was placed

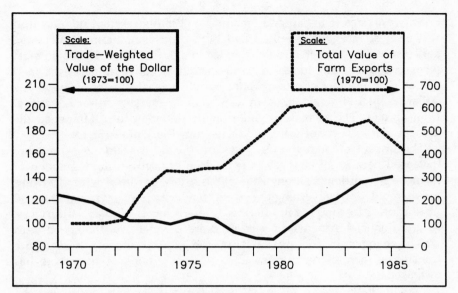

SOURCE: Economic Report of the President 1987.

Figure 1.4 Indexes of the value of the U. S. dollar and of farm exports.

Table 1.4

**Summary of the Farm Sector's Financial Status
(in 1982 Dollars), per Farm, 1940-1985**

Year	Net Farm Income	Farm Assets	Farm Debt	Debt as a Percent of Assets	Interest as a Percent of Earnings
	(- - - thousand dollars - - -)				
1940	5.43	66.38	12.72	19.2	8.8
1945	13.13	109.78	8.43	7.7	2.9
1950	10.11	114.27	9.70	8.5	3.9
1955	8.95	137.25	14.86	10.8	6.3
1960	9.17	172.35	21.41	12.4	9.5
1965	11.37	229.64	35.83	15.6	12.5
1970	11.59	263.12	44.00	16.7	16.7
1975	17.10	385.63	61.35	15.9	18.2
1980	7.74	532.22	87.45	16.4	45.4
1985	11.92	338.86	80.09	23.6	34.5

Source: USDA 1986b.

on interest rates (Duncan and Harrington 1986). The high nominal and real interest rates that resulted had both direct and indirect effects on agriculture. The direct effects included major increases in the interest payments due from indebted farmers and negative impacts on land values. The indirect effects included rapid increases in the value of the dollar as high interest rates in U.S. money markets attracted capital from abroad. The increasing value of the dollar in turn dampened the demand for exports of farm products from the United States.

The impact of high interest rates on the farm sector was probably increased because of changes in federal policies governing regulation of financial institutions. Through the 1980s, there has been a trend toward easing regulations governing the extent to which various types of financial institutions can compete in offering different forms of financial services. As the financial system has become more competitive, the speed with which changes in interest rates in the central money markets are transmitted across the country and to most groups of borrowers has increased.

United States farm policies have also played a role in creating the current conditions. In particular, the 1981 Farm Bill embodied support prices (loan rates) for basic farm commodities that were above market-clearing levels. This encouraged production in the United States, discouraged exports, and encouraged other countries to expand production. Nations that had been substantial importers as well as those that traditionally exported grains and oilseeds registered major increases in their production during the first half of the 1980s. This expanding production capacity is a major reason for the currently depressed state of major farm commodity markets. However, in the Farm Bill of 1985 an effort was made to rectify the problem through provisions calling for a progressive reduction of loan rates to market-clearing levels.

Changes in Farm Structure

Although it can be argued quite effectively that the economic climate confronting agriculture in the 1980s is primarily the product of a series of federal monetary and fiscal policy decisions coupled with national and international economic trends, changes had been occurring within the agricultural sector that had the effect of making it and individuals in it more vulnerable to economic adversity (Raup 1985; Baker 1987). Among these changes were increases in average farm size, increased use of purchased inputs and borrowed capital, and a general reduction in liquidity within the asset structure of the typical farm.

Increasing farm size has been a major factor in increasing production efficiency of American agriculture (U.S. Congress 1986; McKinzie et al.

1987). Farm expansion through increased use of debt capital, however, has also increased the farm's financial vulnerability (Shepard and Collins 1982). As the capital requirements to establish a viable, efficient-sized farming unit have increased, the need for farm families to find ways to gain control of these assets by means other than outright ownership has become increasingly apparent. For example, the average value of farm assets per farm in 1980 was more than $532,000 (Table 1.4). Because few farm families have access to resources of this magnitude at the beginning of their farming careers, they usually must rely on mechanisms, such as leasing or debt financing, to obtain control of part of the resources they will utilize. Debt financing increases the risk of farm failure, however, because a fixed schedule for payment of interest and principal is generally specified and lenders expect payment even during periods of abnormally low income (Ginder 1985). As shown in Table 1.4, the average level of debt per farm more than doubled in the 1960s and almost doubled again in the 1970s.

Information summarized in Table 1.4 also points out the perils associated with debt financing. In particular, the claim that debt service makes on farm income is shown to have increased greatly since 1960. At that time, interest payments amounted to less than 10 percent of total farm earnings (defined here as the sum of net farm income, interest paid, and taxes paid). This percentage rose to about 17 percent in 1970, reflecting both greater capital intensity in agriculture and a slight rise in the sector's debt-to-asset ratio. By 1980, however, the combination of rising interest rates and falling farm income meant that interest payments alone would claim 45 percent of the sector's earnings. Heavily leveraged operators would probably find it very difficult to meet their schedule of interest and principal payments, as well as paying taxes and providing for even minimal family living expenses, under the economic conditions of the 1980s.

Another factor contributing to the vulnerability of the farm sector has been its increasing reliance on purchased inputs. Since 1950, for example, labor has fallen from 38 percent of total farm inputs to about 13 percent, while purchased inputs (such as seed, fuel, fertilizer, feed, chemicals, and interest paid) have increased from 45 percent to 62 percent (Raup 1985). The significance of this change is that most of the labor for the farm sector has traditionally been supplied by owner-operators and their families, while much of the land has either been owned by the family or rented under a share-lease arrangement (which incorporates risk sharing between farmer and land owner). Farm families were able to respond to past periods of economic stress largely by accepting lower returns to their labor and capital, but the increased reliance on purchased inputs (for which prices and payment terms are generally not flexible) reduces the shock-absorbing capacity of the sector (Raup 1985; Baker 1987).

Another trend in the financial structure of the farm sector, which increased its economic vulnerability, was the reduction of liquid assets as a percent of total assets. In 1950, for instance, farmers' holdings of cash (deposits and currency) and U.S. savings bonds amounted to 10 percent of total assets, but by 1982 they had declined to 2 percent (Baker 1987).

These, then, were some of the sectorwide changes that increased the vulnerability of the farm sector to the economic stress imposed by the conditions of the 1980s. However, as noted earlier, the current condition of the farm sector also indicates considerable variation in the situation in which individual farmers find themselves. Some have been able to weather the challenging times much better than others. It is worth noting here, then, some factors that might be likely to render one farm or farm family more vulnerable than another.

Generally speaking, a farmer's vulnerability to economic difficulty under the conditions prevailing during the 1980s is directly related to his debt level. During a period of depressed prices and record-high interest rates, the level of returns to resources has generally not been adequate to allow highly leveraged operators to meet debt service commitments. These heavily indebted operators generally have been placed in this position by one, or a combination, of the following factors:

• Stage in Life Cycle—Beginning farmers are often vulnerable during the early years of their farm operation. Part of the uniqueness of the family farm is that most families accumulate most of their equity capital for the farm from earnings. Thus, the typical family farm life cycle shows the young family beginning to farm with a limited equity and with a relatively high proportion of debt capital and leased or rented resources. Over time, they attempt to reinvest part of their earnings, retiring some or all of their debt, building savings, and expanding their asset base (Boehlje and Eidman 1983). Many have the goal of repaying all debt before retirement, but they tend to be vulnerable during the early stages of their farming careers.

• Farm Expansion—The economics of farming in recent decades has encouraged the continuation of family operations with ownership and management passing to the next generation. Because the farm is then likely for a time to have a greater labor supply and a demand to provide income for two families, there is often an incentive to expand the scale of the farm operation at the time when a family member enters the business. But farm operators who expanded in the 1970s and early 1980s in order to provide an opportunity for a family member often found that their increased debt load was unmanageable.

In similar manner, many producers who sought to expand to gain greater efficiency—for example, adding more land to better utilize available equipment, or acquiring new, larger capacity machinery in order to farm more land—often found that the timing of their expansion moves had placed them in a highly leveraged position at a very inopportune time.

• Unusual Weather or Market Conditions—Producers who encountered adverse weather conditions and suffered loss of part or all of a crop (as occurred in parts of the Corn Belt in 1983 and in the southeastern states in 1986) or who were involved in enterprises that encountered particularly unfavorable market conditions (such as cattle and hog enterprises during much of the early 1980s) have been particularly vulnerable.

Conclusions

Information presented in this chapter clearly indicates that the U.S. farm sector has undergone a considerable financial restructuring over the past few years and that additional financial and other adjustments are likely in the coming years. Further, these adjustments are likely to affect not only farm families but also farm lenders and other segments of the rural community. Thus, we have seen that after a period of favorable economic conditions and rapid expansion in the 1970s, U.S. agriculture entered the decade of the 1980s with substantially increased production capacity, land values, and debt loads. Thereafter, a series of economic changes occurred that were very unfavorable to agriculture and other capital-intensive and export-oriented industries. Reduced export demand resulted in decreased prices for such major commodities as grains and oilseeds, and record-high interest rates made it very difficult for many farmers and ranchers to meet their scheduled payments. Revised expectations of future conditions in agriculture led to sharp reductions in land values, and agricultural lenders were forced to reassess the security of many of their loans. The process of rationalizing the farm sector to new economic expectations is well underway and is likely to continue for several years.

In assessing the events of the last few years, several conclusions seem evident. One is that economic forces and policy decisions external to the farm sector have been primarily responsible both for the expansion that occurred during the 1970s and the subsequent contraction. Thus, a national and international climate featuring strong economic growth, substantial inflation, low real interest rates, and a declining value of the dollar fueled rapid growth in farm exports and, subsequently, expansion of agricultural production capacity and bidding up of land prices. Then in 1979 strong policies were enacted by the United States and other countries to control inflation. Enactment of a tight money policy by the Federal Reserve Board triggered a series of events that brought a reversal of the economic patterns of the 1970s. The U.S. fiscal policies resulted in large budget deficits and subsequently accentuated the pattern of high real interest rates and a highly valued U.S. dollar, both of which worked strongly to the disfavor of the capital-intensive, export-oriented farm sector.

At the same time, it is clear that changes were occurring within the farm sector that increased its vulnerability to the external forces that overtook it in the early 1980s. Growing farm size, which increased the need for reliance on borrowed funds, made the sector more vulnerable to rising interest rates, and the farm sector's growing dependence on purchased inputs increased the potential volatility of net incomes in the face of commodity price fluctuations. Further, trends toward reduced liquidity within the farm asset structure weakened the sector's shock-absorbing ability.

Finally, within the farm sector, circumstances tended to make certain groups of farmers more vulnerable than others. Those who had begun farming during the 1970s generally had had less opportunity to build up a cushion of equity capital and so were typically the first to feel severe financial stress. Similarly, farmers who had chosen to expand substantially during this period or who sustained unusual weather- or market-related reverses in their enterprises found themselves poorly positioned to bear the brunt of the coming crisis.

2

The Structural Characteristics of U.S. Agriculture: Historical Patterns and Precursors of Producers' Adaptations to the Crisis

Don E. Albrecht and Steve H. Murdock

In order to understand the impacts of the farm crisis, it is essential to examine the structural characteristics of agriculture at the onset of the current crisis. The size of the average farm enterprise (in dollars of production or in acres), the form of agriculture being pursued (e.g., crop or livestock production), the level of involvement in nonfarm work activities, the form of farm ownership, and similar characteristics have changed over the last several decades as the farm enterprise has had to respond to changing market conditions and changing forms of technology. Only by knowing the changes and the long-term trends that have occurred in the farm enterprise is it possible to discern the likely effects of the crisis on farm operations and operators.

The purpose of this chapter is to trace the historical patterns of change in agriculture in the United States. The intent is to provide a context for understanding the trends and effects delineated in the remainder of the text. Because it is clearly impossible to describe the patterns of change in all components of the structure of agriculture, only patterns for selected factors are presented here. Those selected have been widely identified as among the most significant in shaping the structure of agriculture in the United States (Cochrane 1979; Schlebecker 1975) and those deemed to be most crucial to determining producers' potential to respond to the crisis (Albrecht and Murdock 1987). Specifically, the factors to be examined include changes in the number and size of farms, the levels of farm concentration and part-time farming, the types of farm organizations, farm tenure patterns, and the level and prevalence of different types of farm labor. These factors are described as they have been affected by the changing technological base in agriculture. The discussion that follows is thus largely a description of agriculture in the United States in the early 1980s.

The Number and Size of Farms

By 1900, most of the potential farmland in the United States was being used for farm production. However, between 1900 and 1910, both the number of farms and the farm population experienced substantial increases as more remote and marginal land was brought into production by an increasing population of rural settlers.[1] By 1910 there were more than 6 million farms, and the farm population surpassed 30 million people. These figures remained virtually constant through the 1940 Census. During this 30-year period the size of the average farm increased as more land was brought into farm production (Table 2.1). Also, while the farm population remained stable through these years, the total population in the United States was increasing. This meant that the proportion of the population living on farms steadily decreased. By 1940, slightly less than one-fourth (23.2 percent) of the U.S. population was living on farms (Table 2.2).

The period since 1940 has been one of rapid change in American agriculture. Technological developments in agriculture have progressively replaced human labor in the production process, and have made it possible for an individual farm operator to cultivate many times the number of acres cultivated by earlier producers (Paarlberg 1980). Due to technological developments such as the all-purpose tractor (Bertrand 1978), by 1982 the average farmer produced enough food and fiber products for 77 people. As a result of such changes, the average farm in the United States was 440 acres in 1982, three times larger than the average farm in 1910. This increase in farm size was a result of the consolidation of farms into larger and larger units, which also corresponded to a rapid decline in the number of farms. By 1982 there were only about 2.2 million farms in the United States, a 65 percent decline from 1940.

The declining number of farms in the United States has also led to a rapid decline in the farm population. The farm population declined from 30.5 million in 1940 to 23.0 million in 1950, to 15.6 million in 1960, to 9.7 million in 1970, to 6.0 million in 1980, and finally to 5.3 million in 1985 (Kalbacher and DeAre 1986). This represents an 82 percent decline in the farm population in a period of 45 years. Once a majority of the U.S. population, farmers comprised only 2.2 percent of the population in 1985. The decline of the farm population is a function of declining family sizes as well as a declining number of

Table 2.1

**Changes in Selected Farming Characteristics
in the United States, 1900–1982**

Year	Number of Farms (1,000)	Size (1,000 Acres)	Percent of Land in Farms	Average Farm Size (Acres)
1900	5,740	841,202	37.0	147
1910	6,366	881,431	38.8	139
1920	6,454	958,677	42.2	149
1930	6,295	990,112	43.6	157
1940	6,102	1,065,114	46.8	175
1950	5,388	1,161,420	51.1	216
1959*	3,711	1,123,508	49.5	303
1969	2,730	1,063,346	47.0	390
1982	2,241	986,797	43.6	440

*Data for Alaska and Hawaii are included for the first time.

Source: Data from 1900 to 1969 are obtained from **Historical Statistics of the United States: Colonial Times to 1970** (U.S. Department of Commerce, Bureau of the Census, 1975). The 1982 data are from the **1982 Census of Agriculture** (U.S. Department of Commerce, Bureau of the Census, 1984a).

Table 2.2

**The Farm Population in the United States,
1900–1985**

Year	Farm Population (1,000)	Percent of Total U.S. Population
1900	29,875	41.9
1910	32,077	34.9
1920	31,974	30.1
1930	30,529	24.9
1940	30,547	23.2
1950	23,048	15.3
1960	15,635	8.7
1970	9,712	4.8
1980	6,051	2.7
1985	5,355	2.2

Source: Data from 1900–1910 are from **Historical Statistics of the United States: Colonial Times to 1970** (U.S. Department of Commerce, Bureau of the Census, 1975). Data for 1920–1985 are from Kalbacher and DeAre, 1986.

Table 2.3

**Number of Farms and Total Farm Sales by
Farm Sales Categories, 1982**

Total Farm Sales Categories	Number of Farms	Total Sales	Percent of Farms	Percent of Total Sales
Less than $5,000	814,535	$1,557,326	36.4	1.2
$5,000-$9,999	281,802	$2,008,511	12.6	1.5
$10,000-$19,999	259,007	$3,694,306	11.6	2.8
$20,000-$39,999	248,825	$7,142,112	11.1	5.4
$40,000-$99,999	332,751	$21,641,796	14.8	16.5
$100,000-$499,999	274,580	$52,781,375	12.3	40.1
$500,000 or more	27,800	$42,764,189	1.2	32.5
Total	2,239,300	$131,589,189	100.0	100.0

Source: **1982 Census of Agriculture** (U.S. Department of Commerce, Bureau of the Census, 1984a).

Table 2.4

**A Comparison of the Percent of Farms and Percent of
Land in Farms by Size Categories, 1900-1982**

Farm Size	1900	1940	1969	1982
10 acres or less				
Percent of farms	4.7	8.3	5.9	8.3
Percent of land in farms	0.2	0.3	0.1	0.1
10-49 acres				
Percent of farms	29.0	29.2	17.4	20.1
Percent of land in farms	5.6	4.5	1.2	1.3
50-99 acres				
Percent of farms	23.8	21.1	16.9	15.4
Percent of land in farms	11.7	8.8	3.2	1.8
100-499 acres				
Percent of farms	39.9	37.0	46.4	39.9
Percent of land in farms	50.6	41.6	27.2	21.4
500-999 acres				
Percent of farms	1.8	2.7	7.9	9.1
Percent of land in farms	8.1	10.5	13.9	14.4
1,000 or more acres				
Percent of farms	0.8	1.7	5.5	7.2
Percent of land in farms	23.8	34.3	54.4	61.0

Source: Data for 1900-1969 are from **Historical Statistics of the United States: Colonial Times to 1970** (U.S. Department of Commerce, Bureau of the Census, 1975). Data from 1982 are from the **1982 Census of Agriculture** (U.S. Department of Commerce, Bureau of the Census, 1984a).

farms. For example, in 1910 there was an average of 5.04 persons per farm. By 1982 this number had declined to 2.41.

Farm Concentration

In 1982 only 1.2 percent of the farms in the United States had gross farm sales of $500,000 or greater, yet these farms accounted for nearly one-third of the nation's total farm sales (Table 2.3). Only about 13.5 percent of the farms had gross sales of $100,000 or more in 1982, but these farms had nearly three-fourths of the farm sales. At the opposite extreme, almost half (49 percent) of the farms in the United States in 1982 had gross farm sales of less than $10,000. These farms, however, account for less than 3 percent of the total farm sales (Table 2.3). Obviously, production in American agriculture is extremely concentrated; and these levels of concentration have increased rapidly in recent years.

An examination of Table 2.4 shows that in 1900 the medium-sized (50–499 acres) farm was dominant in American agriculture. At that time, 64 percent of the farms were between 50 and 499 acres, and these farms accounted for 62 percent of the land in farms. Over one-half of the farmland acreage was in farms between 100 and 499 acres in size. Even in 1900, however, there was evidence of some farm concentration; less than 1 percent of the farms were 1,000 acres or larger, but these farms controlled nearly one-fourth (23.8 percent) of the land in farms.

By 1940, the trend toward larger farms and greater concentration was evident. The percent of the land in farms held by farms between 50 and 499 acres decreased from 62 percent to 50 percent, while the proportion of the land in farms held by farms of 1,000 or more acres increased to 34 percent (Table 2.4). The trend toward larger farms accelerated between 1940 and 1969. By 1969, over half of the land in farms was in farms of 1,000 or more acres, while the proportion of the land controlled by farms with 50 to 499 acres declined to 30 percent (Table 2.4). Between 1969 and 1982, the number of very small and very large farms increased, while the trend toward a decreasing number of medium-sized farms continued. By 1982, the structure of American agriculture could be described as dualistic. At one extreme, there was a large and growing number of small farms that were primarily part-time, hobby, or recreational farms. At the other end of the continuum was a smaller but growing number of large-scale, commercialized farms (Green and Heffernan 1984; Paarlberg 1980; Stockdale 1982). In 1982, farms with 1,000 or more acres controlled 61 percent of the land in farms. Continuing to decline were the medium-sized farms of 50 to 499 acres.

Part-time Farming

At one time in this country, the farm was the sole source of income for most farmers and their families. The 1940 Census of Agriculture, for example, estimated that about 15 percent of all farm operators had 100 or more days of off-farm employment. Typically, off-farm work was considered to be a temporary condition for those involved. It was reserved for those trying to accumulate capital and skills for entrance into farming on a full-time basis or as a mechanism for easing the exit of retiring or marginal producers from agriculture (Albrecht and Murdock 1984; Heffernan et al. 1981).

Today it is becoming increasingly apparent that part-time farming is a stable component of the farm structure and a relatively permanent lifestyle (Paarlberg 1980). In recent decades the prevalence of part-time farming has steadily increased. As shown in Figure 2.1, the proportion of farm operators with 100 or more days of off-farm employment increased from 15 percent in 1940 to 43 percent in 1982. Not only is the proportion of the farm population who have off-farm jobs larger, but those who work off the farm do so for longer periods of time. In addition, female members of farm families are becoming increasingly important in the nonfarm labor force (Coughenour and Swanson 1983; Maret and Copp 1982). It is estimated that 92 percent of the farm families in the United States had some type of nonfarm income in 1977 (Carlin and Ghelfi 1979). Off-farm employment increased in importance until, in 1977, only 39 percent of the income of farm persons came from the marketing of crops and livestock (Paarlberg 1980).

Although the importance of part-time farming has increased throughout the country, it is more prevalent in some parts of the country than in others and is especially important among particular segments of the farm population (Leistritz et al. 1985a). In particular, part-time farming is most common in the South and least common in the Great Plains (Carlin and Ghelfi 1979; Leistritz et al. 1985a). Not surprisingly, the prevalence of part-time farming is much more pronounced among farmers operating small farms than among those operating larger farms. Generally, small farmers have more time available for off-farm work and also have a greater need for additional income to supplement their farm earnings.

In addition, a recent analysis by Albrecht and Murdock (1984) has found that part-time farming is least common in areas where available resources are most conducive to agricultural production and where the employment structure lacks diversity (see also Swanson and Busch 1985; Albrecht and Murdock 1985a). Many farmers apparently become part-time farmers where it is difficult to make a living in agriculture and where off-farm employment opportunities are available. Recent analyses

35

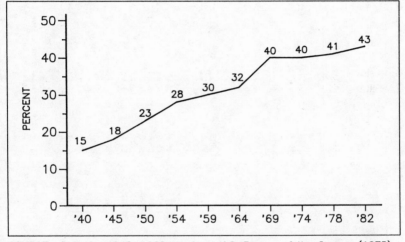

SOURCE: Data for 1940–1969 are from U.S. Bureau of the Census (1975).
Data for 1975–1982 are from U.S. Bureau of the Census (1984a).

Figure 2.1   Percentage of farm operators with 100 or more days
of off–farm employment, 1940–1982.

Table 2.5

Type of Farm Organization by Selected Farm Characteristics for the United States, 1982

Selected Farm Characteristics	Sales ($1,000,000)	Type of Farm Organization			
		Family	Partnership	Family-Held Corporation	Nonfamily-Held Corporation
Farms					
Number		1,945,639	223,274	52,652	7,140
Percent[1]		86.9	9.7	2.4	0.3
Average size of farm (acres)		330	680	2,143	2,024
Land in farms (1,000 acres)		642,380	151,860	112,858	14,451
Percent		68.9	16.3	21.1	1.6
Value of agricultural products sold					
Total ($1,000,000)		77,907	21,520	22,902	8,578
Percent		59.2	16.4	17.4	6.5
Average per farm ($1,000)		40.0	96.4	435.0	1,201.5
Commodities		%	%	%	%
Total cattle and calves	$31,516	48.6	13.9	24.0	13.0
Calves only	3,739	76.6	14.0	7.3	1.0
Fattened cattle only[2]	17,047	29.8	12.1	35.1	22.7
Poultry and poultry products	9,788	59.2	10.2	20.4	10.0
Grain	36,378	72.4	16.4	10.0	0.1
Nursery and greenhouse products	3,820	22.4	8.4	51.2	17.9
Fruit, nuts, and berries	5,842	36.1	20.3	29.0	13.2
Vegetables, sweet corn, and melons	4,135	33.4	20.7	34.1	11.6
Dairy products	16,265	68.7	21.8	8.3	0.8
Hogs and pigs	9,841	69.3	17.1	10.3	2.9
Cotton	3,220	50.8	23.7	19.4	5.3

[1]Percents do not total 100 because some types of farms (such as estates or trusts) are not included.
[2]Fattened cattle are defined as cattle fattened on grain or concentrates for 30 days or more and sold for slaughter.

Source: **1982 Census of Agriculture** (U.S. Department of Commerce, Bureau of the Census, 1984a).

have also shown that off-farm employment may provide farm families with the economic stability to survive during periods of financial stress (Breimyer 1977; Leistritz et al. 1986a), and thus the ability of an area to retain a relatively stable farm population may be dependent on the ability of the area to provide farm residents with off-farm work.

Business Organization of Farming

The family farm has always been proclaimed as the cornerstone of American agriculture. Historically, nearly all farms have been operations for which the family provided most of the management, labor, and capital (Paarlberg 1980). The technological revolution in American agriculture in recent decades, however, has enabled some farms to become very large and highly specialized and some commodities to be produced in an assembly-line fashion (Albrecht and Ladewig 1982). This potential has encouraged many corporations, including some corporations that specialize in nonfarm enterprises, to enter agriculture. Some observers have concluded that this will result in "factories in the field" and in the decline and eventual replacement of the family farm (Barnes and Casalino 1972; Hightower 1971). Concern with corporate farming is rather recent, and a question about the organization of farming was not included in the Census of Agriculture until 1969.

Table 2.5 presents data from the 1982 Census of Agriculture on the characteristics of farms by the type of organization. This table shows that, of the more than 2.2 million farms, the large majority were family farms. Of the remainder, 9.7 percent were partnerships, and 2.7 percent were corporate farms. However, nearly all of the corporate farms (88.1 percent) were family held. These family-held corporate farms are very similar to large family farms except in their legal arrangements (Riemund 1979). In most cases, family-held corporate farms rely on the capital, labor, and managerial decisions of the family in a manner similar to the typical large family farm. In most cases the decision to incorporate was made because of tax or inheritance reasons.

Although family farms remain dominant, there has been an increase in the number of corporate farms. The 1969 Census of Agriculture estimated that there were 21,513 corporate farms in the United States. By 1978 this number had increased to 50,231, and in 1982 there were 59,792 corporate farms, an increase of 178 percent since 1969. This happened during a time period when the total number of farms decreased by 18 percent. Family and nonfamily corporations were not differentiated until 1978. In 1978 there were 5,818 nonfamily corporate farms in the United States. This number increased to 7,140 by 1982.

Although small in number, corporations make a substantial contribution to American agriculture when viewed in terms of the amount of land in farms and value of agricultural products sold. Corporate farms (average size of 2,129 acres) were much larger on the average than were family farms (average size of 330 acres) in 1982. Thus, corporations comprised 2.7 percent of the total number of farms but controlled about 14 percent of the total land farmed. Nearly 89 percent of the land in corporate farms was held by family corporations, however.

The impact of corporate farms on agriculture becomes even more evident when the market value of agricultural products is examined. In 1982 sales from corporate farms constituted 24 percent of the total value of agricultural products sold in this country. In comparison, family farms contributed over 59 percent of the total farm sales. About 6.5 percent of the total farm sales came from nonfamily corporate farms, but nearly three-fourths of corporate farm sales were from family-held corporations. Sales from the average corporate farm totaled more than $526,000 in 1982 compared to about $40,000 from the average family farm. The average nonfamily corporate farm had agricultural sales totaling more than one million dollars in 1982.

Thus, although corporate farms have become an important part of American agriculture, the role of nonfamily corporate farms remains limited. The importance of corporate farms is not uniformly distributed across all agricultural commodities. In 1982, 92 percent of corporate farm cash receipts were from the sale of 9 commodities. The most important commodity produced by corporate farms (as measured by percent of sales) was cattle and calves. This involvement, however, was primarily limited to fattening feedlot cattle rather than calf production (Table 2.5). In fact, nearly one-half of the cash receipts received by nonfamily corporate farms in 1982 came from the marketing of fattened cattle.

Of the other commodities listed in Table 2.5, corporate-farm sales surpassed family-farm sales for only three commodities—nursery and greenhouse products; fruits, nuts and berries; and vegetables, sweet corn, and melons, while the production of cotton, poultry products, grain, dairy products, and hogs and pigs was largely dependent on family farms.

Farm Tenure

The tenant farmer[2] has played an important historical role in American agriculture. Table 2.6 presents information showing the number of farms, percent of farms, and percent of land in farms by tenure status from 1900 to 1982. This table shows that in 1900 about 1 in every 3 American farms was operated by a tenant farmer. For several decades the prevalence of tenant farms steadily increased to 37 percent in 1910, 38 percent in 1920, and finally to 42 percent in 1930. With these high and increasing rates, tenancy issues became one of the major concerns of U.S. agricultural researchers a half-century ago (Brunner and Kolb 1933; Gee 1942; Harris 1941). Researchers and policy makers of the day expressed concern that the American farmer, like the European farmer, would become a landless peasant (Kolb and Brunner 1935; Schmieder 1941).

Since 1935, however, there has been a steady decline in the number and proportion of tenant farms (Kloppenburg and Geisler 1985; Neal and Jones 1950). Their prevalence decreased from 39 percent in 1940 to 12 percent in 1982. Overall, a 91 percent decline in the number of tenant farms has occurred since 1935 (Hottel and Harrington 1979). This has included the almost total elimination of the Southern share-cropper, who was generally considered to be the most disadvantaged tenant farmer (Johnson et al. 1935; Mandle 1983). In 1930, sharecroppers comprised 29 percent of all tenant farmers and 43 percent of the tenant farmers in the South. With mechanization, the labor provided by share-croppers was increasingly less important, and many were forced from the land. In 1959, the census reported the number of sharecropper farms for the last time. Their numbers had become so few that their contributions to agricultural production in the United States had become insignificant.

To a large extent, the decline in the proportion of tenant farms has been accompanied by a corresponding increase in the proportion of part-owner farms. In 1900 only about 8 percent of the farms were part-owner farms, but by 1982 this proportion had increased to 29 percent (Table 2.6). In 1982 part-owner farms were substantially larger on the average than either full-owner or tenant farms. At that time the average part-owner farm was 809 acres, compared to 258 acres for the average full-owner farm and 439 acres for the average tenant farm.

Farm Labor

Historically, American agriculture developed under conditions of plentiful land and scarce labor (Dorner 1983). Thus, securing and maintaining an adequate work force has long been a major problem faced by farm producers (Pfeffer 1983). These labor problems in agriculture are exacerbated by the unique nature of farm production. In contrast to other parts of the American economy, farm production takes place sequentially as opposed to simultaneously (Brewster 1950). In a typical factory operation, for example, production flows through a series of stages, all of which can proceed simultaneously at spatially separated points. In contrast, farm production consists of stages that are typically separated

Table 2.6

Number of Farms, Percent of Farms, and Percent of Land in Farms by Tenure of Farm Operator, 1900-1982

	Full-Owners			Part-Owners			Tenants		
Year	Number of Farms	Percent of Farms	Percent of Land in Farms	Number of Farms	Percent of Farms	Percent of Land in Farms	Number of Farms	Percent of Farms	Percent of Land in Farms
1900	3,202,643	55.8[a]	51.4	451,515	7.9	14.9	2,026,286	35.3	23.3
1910	3,355,731	52.7	52.9	593,954	9.3	15.2	2,357,784	37.0	25.8
1920	3,368,146	52.2	48.3	558,708	8.7	18.4	2,458,554	38.1	27.7
1930	2,913,052	46.3	37.6	657,109	10.4	24.9	2,668,811	42.4	31.0
1940	3,085,491	50.6	35.9	615,502	10.1	28.2	2,364,923	38.8	29.4
1950	3,091,666	57.4	36.1	825,670	15.3	36.4	1,447,455	26.9	18.3
1959	2,116,594	57.8	30.8	834,470	22.5	44.8	735,849	19.8	14.5
1969	1,705,720	62.5	35.3	671,607	24.6	51.8	352,923	12.9	12.9
1982	1,325,773	59.2	34.7	656,249	29.2	53.8	258,954	11.6	11.5

[a]Percents do not total to 100 between 1900 and 1959 because managers are excluded.

Source: Data for 1900-1969 are from **Historical Statistics of the United States: Colonial Times to 1970** (U.S. Department of Commerce, Bureau of the Census, 1975). Data for 1982 are from the **1982 Census of Agriculture** (U.S. Department of Agriculture, Bureau of the Census, 1984a).

by waiting periods because the biological processes involved take time to complete (Madden 1967). Further, unlike other industries where commodities are produced continuously throughout the year, crop production is seasonal.

These aspects of farming create several serious labor management problems. First, farmers need extensive amounts of labor at some points and relatively little at others. Employers are unlikely to be willing to pay workers during periods of inactivity, but if workers are not paid, they will seek alternative employment. Thus, farmers have a problem of securing a sufficient work force during the production cycle and from year to year (Pfeffer 1983). Although a variety of solutions have been tried in respect to obtaining adequate farm labor, the adoption of labor-saving technology has been the emphasis in American agriculture (Berardi and Geisler 1984). Mechanization thus represents an alternative to farm labor and has displaced millions of farm workers. Mechanization has also resulted in larger and fewer farms in this century as producers themselves have been forced from farming. Table 2.7 presents information showing the amount of agricultural employment from 1910 to 1982 in the United States. This table shows a steady decline in total farm employment throughout this century. These declines are especially prominent for family workers. Although there was a 54 percent decline in the number of hired farm workers from 1910 to 1982, the number of family workers declined by 81 percent.

Conclusions

The data in this chapter clearly show that by the beginning of the current farm crisis American agriculture had already made extensive adjustments. Thus, from 1900 to 1982 the number of farms had declined by nearly 3.5 million (61 percent) to 2.2 million farms in 1982, and the farm population declined from nearly 30 million to 5.4 million (a decline of more than 24 million or 82 percent). The average farm's size increased by nearly three times from 1900 to 1982, to an average of 440 acres by 1982. Although the number of small farms has increased in recent years, total agricultural production became increasingly concentrated— 27,000 farms with over $500,000 in sales (1.2 percent of all farms) accounted for nearly one-third (32.5 percent) of the total sales of agricultural products in 1982. By 1982 the United States had developed an increasingly dualistic structure with a large number of small, part-time farms and a small number of large farms that produced the vast majority of the total production of the farm sector.

In addition, by 1982 nearly one-half of the farm operators, were involved in nonfarm employment of some form; 43 percent worked 100

Table 2.7

**Agricultural Employment in Thousands,
1910–1982**

Year	Total Farm Employment[1]	Family Workers[2]	Hired Workers
1910	13,555	10,174	3,381
1920	13,432	10,041	3,391
1930	12,497	9,307	3,190
1940	10,979	8,300	2,679
1950	9,926	7,597	2,329
1960	7,057	5,172	1,885
1970	4,523	3,348	1,175
1982	3,534	1,973	1,561

[1] Based on last-of-the month employment averages.
[2] Includes farm operators and members of their families doing farm work without wages.

Source: Data for 1910–1970 from **Historical Statistics of the United States: Colonial Times to 1970** (U.S. Department of Commerce, Bureau of the Census, 1975). Data for 1982 are from Banks and Mills (1983).

or more days off the farm and more than 90 percent of all farm families received some form of nonfarm income. On the other hand, most farms remained family farms and, except for a few commodities, nonfamily corporate farms did not dominate American agriculture at the onset of the current crisis.

By the onset of the current farm crisis, then, only 2.2 percent of the U.S. population was directly involved in agricultural production, and many of these on only a part-time basis. Farm producers and other Americans had become accustomed to seeing the number of farms decline and rural farm persons being forced to move to urban and rural nonfarm areas. Many of the 2.2 million farms were already marginal in terms of total sales and the capacity for producers to produce anything but partial income for their families from their farm enterprises. In sum, the data in this chapter suggest that, by the onset of the current farm crisis, American agriculture had not only already been involved in more than a half-century of dramatic decline but also had a large number of producers who were vulnerable to any further decline in the economic base of their farm enterprises.

Notes

1. The first agricultural census was taken in 1840 as part of the census of population. From 1840 to 1950, an agricultural census was taken as part of the decennial census. A separate mid-decade census of agriculture was conducted in 1925, 1935, and 1945. From 1954 to 1974, a census of agriculture was taken in the years ending in 4 and 9. Then the census was taken in 1978 and 1982 to adjust the reference year to coincide with the economic censuses. After 1982, the census of agriculture will revert to a 5-year cycle and be taken in years ending in 2 and 7, the same years as the economic censuses. Between 1850 and 1982 the census definition of a farm has changed nine times. Consequently, the data used from the various censuses are not directly comparable. The 1982 definition defines a farm as any place from which $1,000 or more of agricultural products were sold or normally would have been sold during the census year.

2. Broadly defined, a tenant farm is a farm where the operator rents rather than owns the land. Typically, three farm tenure categories are utilized: full-owners own all of the land they operate, tenants rent all of the land they operate, and part-owners own part of the land and rent part of the land they operate. Tenants are by no means a homogeneous class and are generally differentiated on the basis of the degree to which they own and supply the means of production and the amount of control they enjoy over the production process. Cash tenants pay a fixed money rent and provide all nonland inputs and more or less completely determine the manner in which production is carried out. Crop-share and livestock-share tenants return a portion of the animal or vegetable product to the landlord as rent and share with him

management of the operation and the provision of inputs. Sharecroppers provide only labor for a production process that is almost wholly determined by the landlord, who also supplies all means of production (Kloppenburg and Geisler 1985; Pfeffer 1983). In fact, under many states' laws the cropper was officially regarded as a hired worker (Mandle 1983).

3

Demographic, Socioeconomic and Service Characteristics of Rural Areas in the United States: The Human Resource Base for the Response to the Crisis

Steve H. Murdock, Don E. Albrecht, Kenneth Backman, Rita R. Hamm and Lloyd B. Potter

The impacts of the current farm crisis must also be seen in light of the demographic, socioeconomic, and service characteristics of rural areas in the United States as they entered the 1980s. It is this base of people who have had to respond to the crisis. The size of this population, its distribution, and its characteristics, particularly its economic and social characteristics, have placed clear limitations on the magnitude and types of responses to the crisis that have been possible within rural areas. Unless the demographic, socioeconomic, and service characteristics of the population are understood, then, the capabilities of rural areas to respond to the crisis are unlikely to be adequately evaluated.

In this chapter, we present an overview of the history of change in the demographic and socioeconomic characteristics of rural areas in the United States through 1980 and examine its service base in 1980. The changes in the size of the population and in the rural farm, rural nonfarm, and urban components of the population are examined and compared. In addition, its demographic characteristics, such as age, sex, and race or ethnicity, and its socioeconomic characteristics, such as income, education, and occupational patterns, are examined, and the number and relative status of services in areas with small population sizes are compared to those in areas with larger populations. These data do not, of course, adequately account for the variability in the demographic, socioeconomic, or service characteristics of the populations of rural areas experiencing the present crisis, but space limitations do not allow us to examine patterns for areas smaller than the entire United States. Clearly, additional and more detailed analysis must be performed when the impacts on any given group in any given area are being examined. Although limited in detail, the analysis will suggest both the condition of rural areas and the diversity of such conditions at the beginning of the 1980s when the current crisis in agriculture began to be most apparent in rural America.

The Demographic Characteristics of Rural Populations in the United States

Pervasive historical relationships prevail between population change and changes in agriculture (Duncan and Reiss 1956; Brown and Beale 1981). Since the 1940s, counties, and other areas in the United States with high concentrations of their workforces employed in agriculture, have lost population (Brown and Beale 1981). In fact, even during the 1970s when the patterns of renewed rural population growth that came to be known as the nonmetropolitan population turnaround were pervasive, counties with economic bases dependent on agriculture continued to lose population (Bender et al. 1985). For most counties, employment in agriculture has been a major predictor of continued population decline because agricultural technology has replaced the need for agricultural labor and because economies of scale have led to increasing farm consolidation.

Table 3.1 presents a historical view of such changes in the size of the U.S. farm population from 1930 to 1985. As an examination of the data in this table indicates, the farm population changed dramatically from 1930 to 1985. The farm population declined by more than 25 million persons during that period, dropping from nearly 25 percent of the total U.S. population in 1930 to 2.2 percent in 1985. Thus, while the U.S. population was nearly doubling (from 122 million in 1930 to 238 million in 1985), the farm population declined by over 82 percent. This decline has been both large and continuous; each decade since 1940 has witnessed nearly a 25 percent decline in the farm population, and an average of nearly 500,000 persons have left the farm each year from 1930 to 1985. During the 1980s (from 1980 to 1985), the total decline has exceeded 11 percent, and data for 1984 and 1985 suggest that the rate of decline is increasing as a result of the farm crisis (Murdock et al. 1986a).

Tables 3.2 and 3.3 indicate that the rural farm population decline, as measured by the U.S. Census, has occurred despite growth in the total population and in the nonfarm component of the rural population. Thus,

Table 3.1

**Change in the Farm Population of the
United States, 1930–1985**

Year	Farm Population	% Change From Previous Year	% of U.S. Population
1930	30,529,000	----	24.9
1940	30,547,000	0.1	23.2
1950	23,048,000	-24.5	15.3
1960	15,635,000	-32.2	8.7
1970	9,712,000	-27.7	4.8
1980[a]	6,051,000	-37.7	2.7
1981	5,850,000	-3.3	2.6
1982	5,682,000	-3.8	2.4
1983	5,787,000	1.8	2.5
1984	5,754,000	-0.6	2.4
1985	5,355,000	-6.9	2.2
Change 1930–85	-25,174,000	-82.5[b]	---

[a]Beginning in 1980, figures shown use the current farm definition while those before 1980 use the previous farm definitions.
[b]Percent change from 1930 to 1985.

Source: Kalbacher and DeAre, 1986.

Table 3.2

Population and Percent of Population in the United States by Urban, Rural, Rural Farm, and Rural Nonfarm Residence, 1930–1980

| Year | Total Population | Population | | | | Percent of Population | | | |
		Urban	Rural	Rural Farm	Rural Nonfarm	Urban	Rural	Rural Farm	Rural Nonfarm
1930	122,775,046	68,954,823	53,820,223	30,157,513	23,662,710	56.2	43.8	24.5	19.3
1940	131,669,275	74,423,702	57,245,573	30,216,188	27,029,385	56.5	43.5	22.9	20.6
1950[a]	150,697,361	96,467,686	54,229,675	23,048,350	31,181,325	64.0	36.0	15.3	20.7
1960[b]	178,466,732	124,714,055	53,752,677	13,431,791	40,320,886	69.9	30.1	7.5	22.6
1970	203,212,877	149,334,020	53,878,857	10,588,534	43,290,323	73.5	26.5	5.2	21.3
1980[c]	226,545,805	167,054,638	59,491,167	5,617,903	53,873,264	73.7	26.3	2.5	23.8

[a]1950 Census definitions of urban-rural and rural farm and nonfarm
[b]1960 Census definitions of urban-rural and rural farm and nonfarm
[c]1980 Census definitions of urban-rural and rural farm and nonfarm

Source: Data were obtained from the U.S. Department of Commerce, U.S. Bureau of the Census. Vol. 1, "Characteristics of the Population;" Chapter A, "Number of Inhabitants;" Chapter B, "General Population Characteristics;" Chapter C, "General Social and Economic Characteristics;" and Chapter D, "Detailed Population Characteristics." U.S. Censuses of Population and Housing for 1930, 1940, 1950, 1960, 1970 and 1980. Washington D.C.: U.S. Government Printing Office.

Table 3.3

**Percent Change in Population in the United
States by Urban, Rural, Rural Farm, and
Rural Nonfarm Residence, 1930–1980**

| Time
Period | Percent Change in Population | | | | |
	Total Percent Change	Urban	Rural	Rural Farm	Rural NonFarm
1930–1940	7.2	7.9	6.4	0.2	14.2
1940–1950	14.5	29.6	−5.3	−23.7	15.4
1950–1960	18.4	29.3	−0.9	−41.7	29.3
1960–1970	13.9	19.7	0.2	−21.2	7.4
1970–1980	11.5	11.9	10.4	−46.9	24.4
1930–1980	84.5	142.3	10.5	−81.4	127.7

Source: See Table 3.2.

Table 3.4

U.S. Population by 5-Year Age Groups by Urban, Rural, Rural Farm, and Rural Nonfarm Residence: 1950-1980

U.S. Population: 1950

Age	Urban		Rural		Rural Farm		Rural Nonfarm	
	Number	Percent	Number	Percent	Number	Percent	Number	Percent
< 20	29,555,314	30.6	21,543,808	39.7	9,768,841	42.4	11,774,967	37.8
20 - 34	23,947,825	24.8	11,293,270	20.8	4,154,410	18.0	7,118,860	22.9
35 - 49	20,607,533	21.4	9,913,291	18.3	4,198,757	18.2	5,714,534	18.3
50 - 64	14,530,735	15.1	7,036,048	13.0	3,175,826	13.8	3,860,222	12.3
65 +	7,826,279	8.1	4,443,258	8.2	1,750,516	7.6	2,692,742	8.7
Total	96,467,686	100.0	54,229,675	100.0	23,048,350	100.0	31,181,325	100.0

U.S. Population: 1960

Age	Urban		Rural		Rural Farm		Rural Nonfarm	
	Number	Percent	Number	Percent	Number	Percent	Number	Percent
< 20	46,259,061	37.0	22,454,388	41.7	5,639,032	41.9	16,815,356	41.8
20 - 34	24,089,444	19.4	9,328,279	17.4	1,794,272	13.3	7,534,007	18.7
35 - 49	25,019,263	20.0	9,789,368	18.2	2,554,344	18.9	7,266,130	18.1
50 - 64	18,030,631	14.5	7,291,704	13.6	2,217,781	16.5	5,073,923	12.6
65 +	11,315,656	9.1	4,858,243	9.1	1,256,037	9.4	3,601,799	8.8
Total	24,714,055	100.0	53,721,982	100.0	113,461,466	100.0	40,291,215	100.0

U.S. Population: 1970

Age	Urban		Rural		Rural Farm		Rural Nonfarm	
	Number	Percent	Number	Percent	Number	Percent	Number	Percent
< 20	55,611,580	37.3	21,597,069	40.0	4,123,393	39.0	17,473,676	40.3
20 - 34	31,611,878	20.9	9,738,938	18.1	1,465,132	13.8	8,273,806	19.2
35 - 49	26,081,746	17.4	9,134,308	17.0	1,895,805	17.9	7,238,503	16.8
50 - 64	21,759,848	14.6	7,975,636	14.8	1,991,900	18.8	5,983,736	13.7
65 +	14,668,968	9.8	5,432,906	10.1	1,112,304	10.5	4,320,602	10.0
Total	149,334,020	100.0	53,878,857	100.0	10,588,534	100.0	43,290,323	100.0

U.S. Population: 1980

Age	Urban		Rural		Rural Farm		Rural Nonfarm	
	Number	Percent	Number	Percent	Number	Percent	Number	Percent
< 20	51,952,508	31.1	20,463,796	34.4	1,797,696	32.1	18,666,100	34.7
20 - 34	44,775,946	26.8	13,699,242	22.9	958,958	17.0	12,740,284	23.6
35 - 49	26,498,063	15.8	10,164,417	17.1	1,004,835	17.9	9,159,582	17.0
50 - 64	24,827,355	14.9	8,666,092	14.7	1,144,380	20.3	7,521,712	14.0
65 +	19,000,766	11.4	6,497,617	10.9	712,034	12.7	5,785,583	10.7
Total	167,054,638	100.0	59,491,167	100.0	5,617,903	100.0	53,873,264	100.0

Source: See Table 3.2.

the rural nonfarm population has shown steady growth (from 23 million in 1930 to 53 million in 1980) and increased its share of the rural population (from 44 percent in 1930 to 91 percent in 1980). As a result, although the rural nonfarm population was 20 percent smaller than the rural farm population in 1930, by 1980 the rural nonfarm population was nearly 10 times larger than the rural farm population.

The age structure of the rural farm population for 1950, 1960, 1970, and 1980 is shown in Table 3.4. An examination of these data indicates that the age structure of the rural farm population is a bipolar one. For each of the decennial years, the rural farm population had a higher percentage of its population in the age groups of less than 20 and of 65 and over than other population groups. On the other hand, the rural farm population shows the obvious effects of heavy outmigration of young adults, showing a much smaller percentages of adults in the young and middle adult ages, particularly in the 20–34 age group. As a result of such patterns, rural farm populations tend to have high youth and old age dependency ratios (i.e., proportions of persons under age 19 and over age 65 compared to the number 20–64). As shown in Table 3.5, the youth and old age dependency ratios in rural farm populations have consistently exceeded those for urban populations and the old age dependency ratio in the rural farm population has been higher than that for the rural nonfarm population since 1960. The rural farm population then is a population with an age composition indicative of long-term adjustments that have required decreasing amounts of human labor inputs.

The sex composition of the rural farm population also differs from that of other groups. Table 3.6 provides data on the population by sex in the rural, urban, rural farm, and rural nonfarm populations for 1950, 1960, 1970, and 1980. An examination of the data in this table clearly shows rural farm populations to have a disproportionate number of males. For each of the four decades, the rural farm population was over 51 percent male compared to the urban population that was roughly 48 percent male. The rural nonfarm population has shown a pattern of gradual change resembling the rural farm population in terms of its percent male in 1950 and 1960 but developing patterns very similar to the urban population by 1970. The rural farm population is thus a population that is different than urban and rural nonfarm populations in both its sex and its age composition.

As an examination of the data in Table 3.7 indicates, the rural farm and nonfarm populations also differ from the urban population in terms of their racial and ethnic composition. As a result of a rapid decline in its nonwhite population base, the rural farm population had only 111,000 nonwhites in its population base in 1980. From 1950 to 1980 the

Table 3.5

**Youth and Old Age Dependency Ratios for the United States by
Urban, Rural, Rural Farm, and Rural Nonfarm Residence,
1950–1980**

U.S. Dependency Ratio

| | Youth Dependency Ratio | | | | Old Age Dependency Ratio | | | |
Year	Urban	Rural	Rural Farm	Rural Nonfarm	Urban	Rural	Rural Farm	Rural Nonfarm
1950	50.0	76.3	84.7	70.5	13.2	15.7	15.2	16.1
1960	68.9	84.9	86.2	84.5	16.9	18.4	19.2	18.1
1970	70.3	80.4	77.0	81.3	18.6	20.2	20.8	20.1
1980	54.2	62.9	57.8	63.4	19.8	20.0	22.9	19.7

Source: Derived by the authors using sources cited in Table 3.2.

Table 3.6

Population, Percent of Population, and Percent Change in Population by Sex and by Urban, Rural, Rural Farm, and Rural Nonfarm Residence for the United States, 1950–1980

Time Period	U.S. Population							
	Male				Female			
	Urban	Rural	Rural Farm	Rural Nonfarm	Urban	Rural	Rural Farm	Rural Nonfarm
1950	46,891,782	27,941,457	12,078,610	15,862,847	49,575,904	26,288,218	10,969,740	15,318,478
1960	60,422,911	27,414,129	6,978,998	20,435,131	64,291,144	26,338,552	6,482,468	19,856,084
1970	71,945,716	26,944,799	5,403,796	21,541,003	77,388,304	26,934,058	5,184,738	21,749,320
1980	80,292,291	29,760,870	2,918,219	26,842,651	86,758,701	29,733,943	2,699,684	27,034,259
Percent 1950	48.6	51.5	52.4	50.9	51.4	48.5	47.6	49.1
Percent 1960	48.4	51.0	52.0	50.7	51.6	49.0	48.0	49.3
Percent 1970	48.2	50.0	51.0	49.8	51.8	50.0	49.0	50.2
Percent 1980	48.1	50.0	51.9	49.8	51.9	50.0	48.1	50.2
Change 1950–60	28.9	-1.9	-42.2	28.8	29.7	0.2	-40.9	29.6
Change 1960–70	19.1	-1.7	-22.6	5.4	20.4	2.3	-20.0	9.5
Change 1970–80	11.6	10.5	-29.1	24.6	12.1	10.4	-47.9	24.3
Change 1950–80	71.2	6.5	-75.8	69.2	75.0	13.1	-75.4	76.5

Source: See Table 3.2.

Table 3.7

Population, Percent of Population, and Percent Change in Population by Race/Ethnicity and by Urban, Rural, Rural Farm, and Rural Nonfarm Residence for the United States, 1950-1980

U.S. Population

Time Period	Urban			Rural			Rural Farm			Rural Nonfarm		
	White	Black	Other	White	Black	Other	White	Black	Other	White	Black	Other
1950	86,756,435	9,392,608	318,643	48,185,593	5,649,678	394,404	19,715,254	3,158,301	174,795	28,470,339	2,491,377	219,609
1960	110,216,317	13,785,782	711,956	48,244,382	5,051,285	457,014	11,873,087	1,481,971	106,408	36,371,295	3,569,314	350,606
1970	129,069,749	18,338,421	1,925,850	49,037,441	4,211,394	630,022	9,816,142	674,420	97,972	39,221,299	3,526,974	532,050
1980	134,948,026	22,583,845	9,622,767	54,086,986	3,898,504	1,505,677	5,432,353	111,107	74,443	48,654,633	3,787,397	1,431,234
Percent												
1950	89.9	9.8	0.3	88.9	10.4	0.7	85.5	13.7	0.8	91.3	8.0	0.7
1960	88.4	11.1	0.5	89.8	9.4	0.8	88.4	11.0	0.6	90.2	8.9	0.9
1970	86.4	12.3	1.3	91.0	7.8	1.2	92.7	6.4	0.9	90.6	8.2	1.2
1980	80.8	13.5	5.7	90.0	6.6	2.5	96.7	2.0	1.3	90.3	7.0	2.7
Change												
1950-60	27.0	46.8	123.4	0.1	-10.6	15.9	-39.8	-53.1	-39.1	27.8	43.3	59.7
1960-70	17.1	33.0	170.5	1.6	-16.6	37.9	-17.3	-54.5	-7.9	7.8	-0.9	51.8
1970-80	4.6	23.2	400.0	10.3	-7.4	138.9	-44.7	-83.5	-24.0	24.1	7.1	169.0
1950-80	55.5	140.4	2919.9	12.2	-31.0	281.8	-72.4	-96.5	-57.4	70.9	52.0	551.7

Source: See Table 3.2.

nonwhite population on farms declined by over 3 million persons, or by more than 96 percent. Although more than 13 percent of the rural farm population was black in 1950, by 1980 less than 2 percent of that population was black. In 1950 the percent of the rural farm population that was black exceeded that in the urban or rural nonfarm populations, but by 1980 the proportion of the rural farm population that was black was only one-sixth of that in the urban population and one-third of that in the rural nonfarm population. The rural farm population has become an increasingly homogeneous population in terms of its racial and ethnic composition. The rural nonfarm population tends to hold an intermediate position between the farm and urban populations. Its black population increased from 1950 to 1980 but by only one-half the percentage increase in the urban population. As with the rural farm population, however, its composition has tended to become increasingly dominated by the white component. For the rural nonfarm population, however, this has occurred through more rapid growth in its white component than in its black component rather than through the more rapid decline of its black population. Although different in its processual form, the rural nonfarm as well as the farm population, has come to be largely a white population.

Rural farm populations have traditionally had high rates of fertility. As an examination of the data in Table 3.8 suggests, however, rural farm fertility, though still higher than that in other populations, has decreased over the past decades. Whereas the number of children ever born in rural farm populations exceeded that in urban populations by more than 70 percent in 1950, by 1980 the ratio for rural farm populations was roughly 30 percent higher than that in the urban population. In addition, although rural farm fertility was nearly 20 percent higher than that of the rural nonfarm population in 1950, by 1980 the two differed by less than 5 percent.

Mortality in rural areas has consistently been higher than that in urban areas. Thus, numerous scholars have noted the persistence of this differential (Roemer 1976; Navarro 1976; Wright and Lick 1986). Unfortunately, data are not available on mortality patterns for rural farm and rural nonfarm populations. Table 3.9 however, does provide data that show crude death rates for populations in counties with different-sized populations. These data, taken from a recent analysis by Clifford et al. (1986), suggest that mortality remains substantially higher (over 16 percent higher in 1980) in rural than in urban areas. These differences are apparently largely due to differences in the population composition of rural and urban populations. As shown in Table 3.10, when rates in rural areas are standardized by age, race, and sex, rural mortality is little different than urban mortality and, in fact, may be lower for white

Table 3.8

Number of Children Ever Born and Children Ever Born per 1,000 Women by Age of Women and by Urban, Rural, Rural Farm, and Rural Nonfarm Residence for the United States, 1950-1980

			Urban						Rural		
		All Women	Age Groups				All Women	Age Groups			
Children	Year	15-44	15-24	25-34	35-44		15-44	15-24	25-34	35-44	
Number ever born	1950	28,133,925	2,830,530	12,087,648	13,215,747		19,654,657	2,120,220	8,031,130	9,503,307	
	1960	42,588,315	4,414,740	17,785,746	20,387,829		20,441,047	2,121,441	8,262,205	10,057,401	
	1970	48,871,807	4,588,275	19,260,633	25,022,899		19,728,773	1,834,255	7,847,648	10,046,870	
	1980	48,881,650	4,836,249	19,714,387	24,331,014		19,956,014	1,834,118	7,962,076	10,159,820	
Ever born per 1,000 women	1950	1,215	386	1,432	1,792		1,769	537	2,123	2,813	
	1960	1,637	514	2,104	2,269		2,029	602	2,585	3,001	
	1970	1,533	337	2,035	2,840		1,880	434	2,442	3,291	
	1980	1,228	300	1,391	2,561		1,526	372	1,740	2,849	

			Rural Farm						Rural Nonfarm		
		All Women	Age Groups				All Women	Age Groups			
Children	Year	15-44	15-24	25-34	35-44		15-44	15-24	25-34	35-44	
Number ever born	1950	8,687,448	801,120	3,279,693	4,606,635		10,966,209	1,319,100	4,750,437	4,896,672	
	1960	4,898,603	372,436	1,738,066	2,788,101		15,542,444	1,749,005	6,524,139	7,269,300	
	1970	3,657,373	237,149	1,272,379	2,147,845		16,071,400	1,597,106	6,575,269	7,899,025	
	1980	1,658,759	98,606	557,464	1,002,689		18,297,255	1,735,512	7,404,612	9,157,131	
Ever born per 1,000 women	1950	2,074	572	2,378	3,269		1,733	566	2,337	2,494	
	1960	2,133	443	2,798	3,337		1,999	652	2,534	2,889	
	1970	1,946	309	2,567	3,479		1,865	461	2,419	3,244	
	1980	1,595	237	1,911	3,021		1,520	384	1,728	2,832	

Source: Derived by the authors using sources cited in Table 3.2.

Table 3.9

Crude Death Rates (per 1,000) by Residence for the Continental United States, 1970, 1975, and 1980

Year	Most Urban Counties			Residence[1]			Most Rural Counties			Rural/Urban Difference (percent)[2]
	1	2	3	4	5	6	7	8	9	
1970	9.11	8.71	8.77	9.88	11.11	10.40	9.85	11.42	11.68	+23.9
1975	8.63	8.44	8.30	9.19	10.34	9.65	9.18	10.70	10.87	+21.2
1980	8.49	8.39	8.08	8.86	9.82	9.12	8.76	10.09	10.20	+16.4

Note: Crude death rates are three-year averages centered on 1970 and 1975; the 1980 rates are two-year averages for 1979 and 1980.

[1]The residence categories are as follows: (1) core or fringe metropolitan counties with 1 million or more population, (2) metropolitan counties with 500,000 to 999,999 population, (3) metropolitan counties with 50,000 to 499,999 population, (4) nonmetropolitan, adjacent counties with the largest urban place 10,000+, (5) nonmetropolitan, adjacent counties with the largest place 2,500 to 9,999, (6) nonmetropolitan, adjacent counties with the largest place less than 2,500, (7) nonmetropolitan, nonadjacent counties with the largest place 10,000+, (8) nonmetropolitan, nonadjacent counties with the largest place 2,500 to 9,999, and (9) nonmetropolitan nonadjacent counties with the largest place less than 2,500.

[2]The rural/urban percentage difference is calculated by dividing the average difference between the three most rural counties (categories 7–9) and the three metropolitan counties (categories 1–3), by the average of the metropolitan counties.

Source: Clifford et al., 1986.

Table 3.10

Age-Adjusted Death Rates (per 1,000) by Sex, Race, and Residence for the Continental United States, 1970, 1975, and 1980

Year and Race Group	Residence[1]									Rural/Urban Difference[2] (percent)	White/Nonwhite Difference[3] (percent)
	Most Urban Counties						Most Rural Counties				
	1	2	3	4	5	6	7	8	9		
Males											
1970: White	11.89	11.85	11.77	12.01	12.02	11.87	12.19	12.15	11.78	1.7	22.6
1970: Nonwhite	14.81	13.82	15.00	14.85	14.75	13.50	15.68	15.51	13.92	3.4	
1975: White	10.77	10.90	10.82	10.96	11.07	10.97	11.18	11.33	10.89	2.8	22.0
1975: Nonwhite	13.11	12.88	13.35	13.52	13.48	12.82	14.17	14.24	13.11	8.3	
1980: White[4]	10.02	10.06	9.86	9.99	10.10	10.04	10.20	10.25	9.91	1.4	25.7
1980: Nonwhite[4]	12.24	12.28	12.60	12.68	12.84	12.40	13.25	13.20	12.17	4.0	
Females											
1970: White	7.14	6.93	6.87	7.08	6.90	6.87	6.97	6.97	6.84	-0.8	38.0
1970: Nonwhite	9.40	8.96	9.95	9.80	9.41	8.92	10.38	10.14	9.35	5.4	
1975: White	6.33	6.21	6.13	6.17	6.13	6.13	6.19	6.17	6.04	-1.5	33.1
1975: Nonwhite	7.90	8.03	8.32	8.57	8.17	7.87	8.74	8.52	7.77	3.2	
1980: White[4]	5.88	5.78	5.66	5.70	5.58	5.56	5.64	5.59	5.45	-3.7	31.7
1980: Nonwhite[4]	7.30	7.35	7.60	7.43	7.48	7.23	7.73	7.69	7.13	1.4	

[1] The residence categories are: (1) core or fringe metropolitan counties with 1 million or more population, (2) metropolitan counties with 500,000 to 999,999 population, (3) metropolitan counties with 50,000 to 499,999 population, (4) nonmetropolitan, adjacent counties with the largest urban place 10,000+, (5) nonmetropolitan, adjacent counties with the largest place 2,500 to 9,999, (6) nonmetropolitan, adjacent counties with the largest place 10,000+, (7) nonmetropolitan, nonadjacent counties with the largest place 2,500 to 9,999, and (8) nonmetropolitan, nonadjacent counties with the largest place less than 2,500, (9) nonmetropolitan nonadjacent counties with the largest place less than 2,500.

[2] The rural/urban percentage difference is calculated by dividing the average difference between the three most urban counties (categories 7-9) and the three metropolitan counties (categories 1-3) by the average age-adjusted rate of the metropolitan counties.

[3] The white/nonwhite percentage difference is merely the average difference between all residence categories. a percent of the white average for all residence categories.

[4] Adjusted using the 1970 U.S. population as the standard.

Source: Clifford et al., 1986.

females in more rural counties than in more urban counties. As with several other demographic characteristics discussed in this chapter, then, mortality in rural areas appears to be becoming increasingly similar to that in urban areas.

Perhaps no recent pattern of population change in rural areas has received greater attention than the rural population turnaround; that is, the pattern of net inmigration to nonmetropolitan areas from metropolitan areas that occurred in the 1970s after decades of rural population decline involving net losses of migrants from nonmetropolitan to metropolitan areas (Brown and Beale 1981). As for mortality, migration data are not available for rural farm and nonfarm populations, and thus data for metropolitan and nonmetropolitan areas are presented here. The data in Table 3.11 clearly show both the historical patterns of migration between nonmetropolitan and metropolitan areas, the dramatic change in such patterns during the 1970s, and the recent return to historical patterns. Thus, although nonmetropolitan areas gained more than 1.3 million persons due to outmigration from metropolitan areas during the period from 1975 to 1980, since 1982–1983 the return to the historical patterns of net outmigration from nonmetropolitan areas to metropolitan areas is again evident. It appears that the migration turnaround may have been a relatively short-lived phenomena (Richter 1985). In fact, data for 1980 to 1985 show metropolitan areas had a population increase of 5.9 percent, while nonmetropolitan areas increased by only 3.7 percent (Engels 1986). In addition, although no data are available for separate rural farm and nonfarm populations, analyses of patterns in counties with large percentages of their labor forces employed in agriculture (Richter 1985; Bender et al. 1985) suggest that areas dominated by agriculture never shared in the turnaround patterns of net inmigration and that the turnaround was largely a rural nonfarm phenomenon. In sum, it appears that the historical pattern of net outmigration from rural farming areas to urban centers is still occurring.

Taken together, the data in Tables 3.1 through 3.11 provide an overview of the demographic characteristics of the rural population compared to other residence groups. An evaluation of the data in these tables suggests that the rural farm population has declined to roughly one-sixth of its 1930 size, while the rural nonfarm population has more than doubled. Rural farm and nonfarm populations are increasingly concentrated in younger and older age groups, are increasingly racially and ethnically homogeneous, and retain levels of fertility that exceed those in the urban population. They have mortality patterns that are reflective of their more aged population bases and have shown historical patterns of continued net outmigration. Their populations reflect the technological advances and the long decline in labor needs in U.S. agriculture and

Table 3.11

Migration in U.S. Metropolitan and Nonmetropolitan Counties, 1965-1984

County Group/ Period	Inmigrants	Outmigrants	Net Migrants
Metropolitan	(- - -numbers are in thousands- - -)		
1965-1970	5,457	5,809	-352
1970-1975	5,127	6,721	-1,594
1975-1980	5,993	7,337	-1,344
1980-1981	2,156	2,350	-194
1981-1982	2,217	2,366	-149
1982-1983	2,088	2,066	22
1983-1984	2,609	2,258	351
Nonmetropolitan			
1965-1970	5,809	5,457	352
1970-1975	6,721	5,127	1,594
1975-1980	7,337	5,993	1,344
1980-1981	2,350	2,156	194
1981-1982	2,366	2,217	149
1982-1983	2,066	2,088	-22
1983-1984	2,258	2,609	-351

Source: U.S. Department of Commerce, U.S. Bureau of the Census, "Geographical Mobility" **Current Population Report** P-20 Nos. 368, 377, 384, 393, and 407. Washington D.C.: U.S. Government Printing Office, 1981, 1983, 1984b, and 1986.

Table 3.12

Labor Force by Industry and Urban, Rural, Rural Farm, and Rural Nonfarm Residence for the United States, 1950

Industry	Number	Total Percent	Percent Urban	Percent Rural	Percent Rural Farm	Percent Rural Nonfarm
Ag, forestry, fisheries	7,005,403	11.7	1.1	35.4	70.4	8.6
Mining	929,152	1.5	0.8	3.2	1.3	4.6
Construction	3,439,924	5.7	5.6	6.1	3.1	8.4
Manufacturing	14,575,692	24.2	27.2	17.6	9.3	24.0
Transportation, communication and utilities	4,368,302	7.3	8.4	4.8	2.0	7.0
Wholesale trade	1,975,817	3.3	4.0	1.8	0.8	2.5
Retail trade	8,571,752	14.3	16.3	9.7	3.5	14.4
Finance, insurance and real estate	1,916,220	3.2	4.1	1.2	0.5	1.8
Services	5,453,937	9.1	10.3	6.3	2.6	9.1
Professional services	9,349,096	15.6	17.4	11.3	5.3	15.8
Public administration	2,488,778	4.1	4.8	2.6	1.2	3.8
Total (N)	60,074,073	---	41,533,255	18,540,818	8,040,796	10,500,022

Source: See Table 3.2

in other industries in rural trade centers. Their young adult age groups have been depleted due to the heavy outmigration of young adults who are no longer needed to provide the labor and services made superfluous by agricultural and transportation technology. In the rural farm population, the nonwhite base has been depleted as a result of being composed largely of nonowners of land during a period in which expansion in land ownership was the key to survival, and in rural nonfarm areas, the growth of the nonwhite population has failed to keep pace with the white population because of decreasing employment opportunities for nonwhites in nonfarm as well as farm-related jobs. In sum, then, both the rural farm and nonfarm populations have been required to make dramatic adaptations to the rapidly changing circumstances in agriculturally based areas.

Socioeconomic Characteristics of Rural Populations in the United States

In this section the socioeconomic characteristics of the rural farm and nonfarm populations are examined in comparison to the urban population component. The characteristics include employment by industry and occupation and education. The intent of this section is to describe the changing patterns of economic activity in rural farm and nonfarm populations and the implications of these patterns for the social and economic conditions of rural populations.

Tables 3.12 and 3.13 show employment by industry in rural farm and nonfarm and other population components for 1950 and 1980. The results in these tables point to a clear evolution in the economic base of rural farm populations. Over 70 percent of the labor force in rural farm populations in 1950 was engaged in agriculture, forestry and fishing, but by 1980 less than one-half of the labor force living on farms was employed in agriculture. By 1980, 23 percent of the rural farm population was employed in services (including professional services) and nearly 12 percent was engaged in manufacturing. Urban populations show even more rapid transitions to service-based economies and reductions in manufacturing employment. Rural nonfarm populations display patterns similar to those in urban populations. Rural nonfarm populations experienced a shift towards manufacturing-based economies during the 1950s and 1960s, followed by a reduction in manufacturing employment after 1970 and a rapid increase in employment in service industries. It is important to note, however, that the percentage of the labor force engaged in manufacturing in rural nonfarm areas in 1980 exceeded that in urban areas.

Table 3.13

Labor Force by Industry and Urban, Rural, Rural Farm, and Rural Nonfarm Residence for the United States, 1980

Industry	Number	Percent	Labor Force			
			Percent Urban	Percent Rural	Percent Rural Farm	Percent Rural Nonfarm
Ag, Forestry, fisheries	2,913,589	2.5	0.9	7.6	41.1	3.8
Mining	1,028,178	0.9	0.6	1.7	0.8	1.8
Construction	5,739,598	4.9	4.4	6.5	4.0	6.8
Manufacturing	21,914,754	18.7	17.7	21.7	11.9	22.8
Transportation, communication and utilities	7,087,455	6.0	6.1	5.7	3.5	6.0
Wholesale trade	4,217,232	3.6	3.7	3.2	2.7	3.3
Retail trade	15,716,694	13.4	13.8	12.0	6.9	12.6
Finance, insurance and real estate	5,898,059	5.0	5.6	3.3	2.2	3.4
Services	27,976,330	23.7	25.0	20.0	13.7	20.7
Professional services	19,811,819	16.9	17.6	14.6	10.8	15.0
Public administration	5,147,466	4.4	4.6	3.7	2.4	3.8
Total (N)	117,451,174	---	89,497,838	27,953,336	2,772,134	25,181,202

Source: See Table 3.2

Finally, the changing size of the labor force among the different population components is also important to note. Whereas the size of the rural farm labor force was declining from more than 8.0 million in 1950 to 2.8 million by 1980, the labor force in urban areas was increasing from 41.5 million to 89.5 million and that in rural nonfarm areas from 10.5 million to 25.2 million. In addition, although the rural farm labor force declined by over 64 percent in the 30 years from 1950 to 1980, this decline was not sufficient to maintain the dominance of agriculture in the rural farm economy. The number of rural farm residents employed in agriculture declined by 80 percent (from 5.7 million in 1950 to 1.1 million in 1980). Between 1950 and 1980, the number of persons engaged in agriculture in rural farm populations declined from 1.5 million to 1.1 million (a decline of nearly 24 percent, but total employment in the rural farm population decreased by even more—31 percent).

In rural nonfarm populations, employment is also increasingly dependent on nonagricultural activities. It is important to note, however, that the evolution of rural nonfarm economies away from agriculture has largely resulted in their acquiring dependence on forms of manufacturing and service employment that are increasingly vulnerable to declining market conditions similar to those which have affected U.S. agriculture (Miller and Bluestone 1987). The data in Tables 3.12 and 3.13 thus suggest that the economic base of agriculture has not been able to support its human resource base. As a result, even among rural farm populations there has been a clear evolution in the structure of the economic base away from extractive and towards service-based industrial economies, while in rural nonfarm populations dependence has shifted toward manufacturing and services that show increasing vulnerability to declining world market conditions (Miller and Bluestone 1987).

Tables 3.14 and 3.15 display data on occupational patterns in rural and urban populations for 1950 and 1980. In general, these data verify the patterns shown for the industrial data. Over 70 percent of the rural farm labor force in 1950 was employed as farm operators/managers or farm workers, but by 1980 the proportion had decreased to 45 percent. In this group, farm laborers showed the most dramatic decline, from more than 20 percent (in 1950) to only 10 percent of the rural farm labor force by 1980. It is equally important to note that the rural farm and rural nonfarm labor forces have increasingly come to be composed of persons employed in operator, precision production, and service occupations, while urban populations have increasingly become involved in managerial and professional as well as service occupations. It is evident, then, that rural farm and nonfarm populations are both in-

Table 3.14

Labor Force by Occupation and Urban, Rural, Rural Farm, and
Rural Nonfarm Residence for the United States, 1950

Occupation Group	Labor Force					
	Number	Percent	Percent Urban	Percent Rural	Percent Rural Farm	Percent Rural Nonfarm
Managerial occupations	5,076,848	8.8	10.0	6.1	2.0	9.4
Professional/ specialty occupations	4,988,963	8.7	10.1	5.4	2.5	7.7
Technical and support occupations	11,708,022	20.3	21.5	17.8	8.9	24.4
Sales occupations	4,044,251	7.0	8.3	4.2	1.6	6.1
Administrative occupations	7,071,283	12.3	15.4	5.4	2.4	7.7
Service occupations	4,511,677	7.8	9.3	4.5	1.5	6.8
Private household occupations	1,488,388	2.6	2.8	2.0	1.2	2.7
Farm operators and managers	4,322,809	7.5	0.3	23.2	49.8	2.4
Farm workers and laborers	2,514,843	4.4	0.6	12.6	21.3	5.9
Operators, fabricators and Labors	3,765,394	6.5	6.4	7.1	3.8	9.8
Precision production craft and repair	8,162,499	14.1	15.3	11.7	5.0	17.1
Total (N)	57,654,977	----	39,657,765	17,997,212	7,930,868	10,066,344

Source: See Table 3.2.

Table 3.15

**Labor Force by Occupation and Urban, Rural, Rural Farm, and
Rural Nonfarm Residence for the United States, 1980**

			Labor Force			
Occupation Group	Number	Percent	Percent Urban	Percent Rural	Percent Rural Farm	Percent Rural Nonfarm
Managerial occupations	10,133,551	11.0	11.9	8.1	4.8	8.5
Professional/ specialty Occupations	12,018,097	13.0	14.1	9.7	6.8	10.1
technical and support	2.981,951	3.2	3.5	4.8	1.3	5.4
Occupations sales occupations	9,760,157	10.6	11.3	8.3	5.0	8.9
Administrative occupations	16,851,398	18.3	19.9	13.1	9.1	13.2
Service occupations	12,040,073	13.1	13.5	11.4	6.7	11.9
Private household occupations	589,352	0.6	0.6	0.6	0.4	0.6
Farm operators and managers	1,298,670	1.4	0.2	5.0	35.9	1.5
Farm workers and laborers	1,334,123	1.5	0.9	3.2	10.1	2.5
Operators, fabricators and laborers	12,495,844	13.6	11.5	19.6	11.1	20.3
Precision production craft and repair	12,594,175	13.7	12.6	16.2	8.8	17.1
Total (N)	92,097,391	---	69,351,204	22,746,187	2,395,268	20,350,919

Source: See Table 3.2.

Table 3.16

**Median School Years Completed for Persons 25 Years
Old and Older for the United States by Urban, Rural,
Rural Farm, and Rural Nonfarm Residence, 1950—1980**

		Median School Years Completed		
Year	Urban	Rural	Rural Farm	Rural Nonfarm
1950	10.2	8.6	8.8	8.4
1960	11.1	9.2	8.8	9.5
1970	12.2	11.0	10.7	11.2
1980	12.5	12.3	12.3	12.3

Source: See Table 3.2.

creasingly employed in nonagricultural industries as well as nonagricultural occupations.

Table 3.16 presents data on the educational characteristics of rural farm populations. These data suggest increasing similarity in the educational characteristics of rural and urban residents. Whereas the rural farm population lagged behind the urban population by 1.4 years in terms of median years of education in 1950, by 1980 the difference had been reduced to 0.2 years. A similar trend is evident for comparisons between rural farm and rural nonfarm populations. Recent decades have thus seen significant progress in the socioeconomic status of the rural population.

The data in Tables 3.11–3.16 provide additional evidence concerning the adaptation of the rural population to the changing structure of agriculture and other extractive industries. An examination of the data in these tables indicates that rural populations have faced reduced needs for agricultural labor that have not only dramatically reduced the size of their workforces and the proportion of their workforces engaged in agriculture and related industries but also expanded the level of employment in nonagricultural industries, such as manufacturing and services. The data further suggest that by so doing, the rural population has made progress in closing the educational disparities that have traditionally existed between rural and urban populations. An evaluation of these data thus suggests that the adaptation of the rural population has been both substantial in magnitude and at least partially successful in altering the socioeconomic conditions of this population.

Service Characteristics of
Rural Areas in the United States

The service characteristics of rural areas in the United States are difficult to describe because few comprehensive assessments of these characteristics have been made. Rather, data are largely available for only selected jurisdictions, and the measures of service availability and quality vary so widely from one area to another that it is often impossible to compare data for one area to those for other areas. One of the few attempts to assess services available in rural areas throughout the United States was the National Rural Community Facilities Assessment Study (NRCFAS) survey conducted for Farmers Home Administration in 1982 by Abt Associates Inc. (Reid et al. 1984). This random sample survey of 520 rural communities in the 48 contiguous states provides comparable information on the characteristics of selected services in rural communities of different population sizes. In Table 3.17, summary data from this survey are presented that allow the general service characteristics of services in rural areas in the United States to be assessed.

An examination of the data in Table 3.17, clearly show that rural communities in the United States (particularly those with fewer than 2,500 people) and rural areas outside incorporated places tend to have fewer and generally lower quality services than communities with larger populations and the United States as a whole. For example, as shown in this table, rural places of less than 2,500 population and rural areas outside of incorporated places are less likely than areas with larger populations to have public water systems, their water systems are less likely to be provided to all households, their water systems are tested for water contamination less often, they have fewer hospitals, they have fewer specialized medical services such as hemodialysis and psychiatric services, and they are more likely to have roads with uninspected bridges. Although these data are for a single period of time, other analyses (Honadle 1983; Doeksen and Peterson 1987) suggest that rural-urban differences in services were even more disparate in earlier decades and may have been at their highest level of development in the late 1970s and early 1980s. Such analyses also suggest that the 1980s have brought about disproportionate cutbacks in funds for services in rural areas so that service quality has probably declined in rural areas relative to more urban areas during the 1980s (Long et al. 1987).

In sum, although the data available are limited, it appears that rural areas in the United States have fewer and lower quality services than more urban areas. As a result, rural areas had little excess capacity to use in serving persons affected by the crisis and could ill-afford a loss of the service base or fiscal support likely to accompany a loss of population.

Conclusion

The data in this chapter clearly show that the demographic, socioeconomic, and service bases of rural areas had already changed dramatically by the onset of the current crisis. The rural farm population declined to only 5.4 million by 1985 (by 82.5 percent from 1930), while rural nonfarm populations increased. However, both rural farm and nonfarm populations were older and more racially and ethnically homogeneous and had higher rates of fertility and a predominant pattern of outmigration, particularly among young adults. By 1980 less than one-half of rural farm residents listed their major industry or occupation of employment as that of agriculture, and rural farm and nonfarm residents' levels of median education and income were lower than those for urban residents. In sum, the population of rural areas, particularly

Table 3.17

**Selected Characteristics of Services for Rural Communities in 1980
by Population Size in 1978 (Estimated)**

Community Service Characteristic	Population Size in 1978						
	20,000-49,999	10,000-19,999	5,500-9,999	2,500-5,499	Less Than 2,500	Unincorporated Areas	Total U.S.
	(- - - - - - - - percent with characteristic - - - - - - - - - -)						
No public water system, 1980	0	0	0	0.5	14.5	62.5	44.7
Water service for only 1-33% of households, 1980	0	0	0	0.5	14.7	73.9	52.2
Water service to 100% of households, 1980	58.7	68.6	60.0	68.7	62.2	3.3	24.1
Have public water systems in 1980 that did not test for cloriform bacteria at least 12 times last year	0	4.9	1.9	5.7	10.4	20.5	13.9
Served by waste-water treatment plants, 1980	94.3	96.5	96.2	93.1	62.0	13.4	30.4
Served by waste-water treatment plants with ratio of existing flow to design flow over 100%, 1980	11.7	14.0	22.5	22.9	25.4	19.2	22.8
With 5 or more hospital beds, 1977	77.5	63.5	75.5	68.0	67.1	58.6	61.8
With 300 or more hospital beds, 1977	91.9	75.8	83.6	73.4	68.5	65.4	67.2
With 5 or more emergency rooms available, 1977	72.8	55.8	73.3	61.7	59.1	52.9	55.5
With neonatal intensive care services, 1977	55.8	39.5	51.2	42.5	40.5	35.2	37.1
With hemodialysis services, 1977	59.9	50.6	50.0	52.8	46.8	45.0	46.1
With psychiatric services, 1977	74.9	55.9	58.6	59.9	50.2	55.8	54.6
With 5 or more nursing homes available, 1978	92.3	94.3	91.7	94.2	95.7	91.1	92.5
With fire service, but lacking complete hydrant coverage or truck capacity of 3,000 gallons, 1980	11.1	8.3	8.5	11.6	20.5	54.1	41.3
With 51 or more miles of local roads, 1980	95.3	84.6	47.1	13.1	1.3	9.6	9.6
With 26 or more miles of road, 2 lanes, 10 feet wide, 1980	96.2	93.3	83.1	38.4	3.3	21.4	19.6
With no unpaved roads, 1980	60.6	43.9	47.0	49.3	45.8	10.0	22.7
With no roads closed awaiting repairs, 1980	94.1	95.1	98.8	93.3	90.6	33.4	53.7
With no locations made frequently impassable by natural events, 1980	76.6	88.7	76.2	80.3	83.6	20.5	42.3
With roads with no narrow bridges, 1980	77.1	66.3	45.9	41.1	17.7	16.6	19.4
With roads with all bridges inspected in last 3 years, 1980	64.3	67.5	49.3	35.4	20.5	22.6	23.8

Source: J.N. Reid, T.F. Stinson, P. Sullivan, L. Perkinson, M.P. Clarke, and E. Whitehead, **Availability of Selected Public Facilities in Rural Communities: Preliminary Estimates,** Washington, D.C.: ERS Staff Report No. AGES840113, Economic Development Division, Economic Research Service, U.S. Department of Agriculture, 1984.

the rural farm population base, by 1980 was one which clearly reflected long years of adjustment to the declining need for agricultural labor. Demographically, it was a population base that displayed little potential for long-term population growth and retention, and in fact, its age and other characteristics already suggested the likelihood of further decline. Its socioeconomic characteristics suggested that many of its members were already experiencing relative deprivation compared to urban Americans and were likely to be particularly vulnerable to further decline in their income base. By the onset of the current crisis, the population base of rural areas was thus already relatively disadvantaged. Finally, at the onset of the crisis, the service base of rural America was undeveloped relative to other areas in the United States. It had fewer services and lower quality services than urban America.

Taken together, an examination of the data on the population, socioeconomic, and service characteristics of rural America suggests that, at the time the crisis became most severe, rural America was ill-prepared to weather an economic downturn, involving reductions in income, populations, services bases, and fiscal resources. Its reserves had been largely exhausted by decades of decline. It was in fact, an area that we would argue was highly vulnerable, an area for which further decline could lead to largely negative and permanent changes in the quality of life in rural America.

PART TWO

The Characteristics, Impacts and Long-term Implications of the Crisis

4

The Financial Characteristics of Production Units and Producers Experiencing Financial Stress

F. Larry Leistritz and Brenda L. Ekstrom

During the decade of the 1980s American agriculture has been required to adjust to a new set of economic realities. The implications of this adjustment process have been multifaceted, but many observers have summarized the effects of the current economic environment on farmers and ranchers with the phrase *financial stress*. In this chapter, we first define the meaning of financial stress and discuss alternative measures of producers' actual or potential levels of stress. Then we examine the nature and extent of financial stress in American agriculture. Special emphasis is given to describing variations in the incidence of financial stress among farming areas, among farms of different sizes, and among farms producing different types of commodities.

Defining Financial Stress

Although the term *financial stress* has come to be widely used in describing current economic conditions in the farm sector (Petrulis et al. 1987b; Johnson et al. 1987), the meaning of the term has not always been clearly defined. Generally, financial stress results from an inability to meet cash flow commitments, that is, from the inability to meet financial obligations when due (Brake 1983). When a farm household does not have sufficient cash available to meet the cash expenses of the farm operation, family living, and scheduled debt service, they are said to be experiencing financial stress. Sometimes such a cash flow problem can be resolved quite simply, but in other situations it may signal the impending end of a farming career. In order to fully understand a farm's financial situation, then, it is important to examine a number of indicators of income and net worth.

One major determinant of a household's ability to meet financial obligations is the current income level. Both farm income and income from off-farm sources are relevant, and in some cases off-farm income may be the key to meeting financial commitments (Leistritz et al. 1985b; Harrington and Carlin 1987). When current income from all sources is inadequate to meet current expenses, several alternatives for meeting these obligations may still exist. For instance, it may be possible to utilize savings from previous years, to sell selected assets, or to borrow the funds needed to meet current commitments, but the extent to which these options will be practical generally depends on the nature of the family's assets and liabilities and the relationship between them. This relationship is often summarized in a balance sheet or net worth statement.

Examination of the balance sheet can provide insights concerning both the solvency and the liquidity of a farm business or household (Table 4.1). *Solvency* refers to the overall relationship of the farm household's assets to its debts. When the value of the assets exceeds the total amount of the debts, the business is said to be solvent. If the assets were liquidated, the proceeds would be more than adequate to repay all creditors. When the value of the assets exceeds the amount of debt by a wide margin, the surplus represents potential collateral for additional borrowing. Thus, a farm's solvency position provides a measure of its risk-bearing ability because a farm with a substantial net worth or equity (the difference between the value of assets and amount owed to creditors) has the potential to borrow additional funds to meet short-run needs. In fact, during the 1970s many producers used their growing net worth arising from appreciating land values as the basis for expanded borrowing to meet a variety of needs (Ginder 1985; Harrington and Carlin 1987). Conversely, when the amount owed approaches the total value of all assets, the firm's risk-bearing ability is very limited. Lenders are likely to be reluctant to extend further credit and may in fact press the borrower to reduce the level of debt or to provide additional collateral. This situation has confronted many operators during the 1980s as declining land values have eroded their equity.

Liquidity refers to the ability to meet short-term financial obligations. *Liquid assets* is a term frequently used in referring to assets that are in the form of cash or can be readily converted to cash. A farm that has a considerable portion of its assets in liquid form (e.g., savings accounts, grain in storage) is less likely to have problems in meeting unexpected financial needs than one that has almost all of its assets invested in land, machinery, or breeding livestock. While assets of these latter types

Table 4.1

Components of Farm Financial Statements

Financial Statement	What Is Measured	Components
Balance sheet	Solvency and liquidity	**Farm assets (+ nonfarm assets)** **− Farm liabilities (− nonfarm liabilities)** **= Farm business net worth or (farm household net worth)**
Income statement	Profitability	Farm revenue − Farm operating expenses = Net operating margin
		Net operating margin − Interest paid − Depreciation = Net cash farm income
		Net cash farm income (+ nonfarm earnings) +/− Inventory change (+ other nonfarm income) **= Net farm income or (total family income)**
Cash flow statement	Liquidity	Net operating margin + Capital sales (+ nonfarm income) = Total cash in flow
		Total cash in flow − Interest paid − Principal payments (− family living expenses) **= Net cash balance**

Note: Items in parentheses are components that will be included if the focus is on the farm household but excluded if the analysis is for the farm business only.

could possibly be sold to meet short-term financial needs, a forced liquidation of such assets often involves either a substantial sacrifice in price, or a major reduction in the efficiency of the remaining farm unit, or both (Boehlje and Eidman 1983).

Because farming and ranching are known to involve a variety of risks arising from such factors as weather, disease, and changing market conditions, many farm and ranch operators have historically preferred to retain a substantial portion of their assets in liquid form; however, as noted in Chapter 1, the liquidity of the farm sector as a whole declined substantially during the 1970s. Although there is still some disagreement whether this reduction in liquidity resulted primarily from decisions by farmers to invest a higher percentage of their assets in land, machinery, and other less liquid forms or was primarily the product of sharply rising land values (Boehlje and Eidman 1983; Brake 1983), it is clear that the reduction in farm sector liquidity lessened the ability to cope with adverse economic conditions.

Financial stress, then, may arise when market forces drive farm income or profits below their normal levels, but the critical factor in determining whether an individual, firm, or economic sector will experience such stress is its capacity to adjust to adverse economic events. When the adjustments required exceed the capacity to adjust, financial stress occurs.

Several indicators of adjustment capacity useful in determining whether a farm household will experience financial stress have been identified (Jolly et al. 1985). The financial health of a farm business can be assessed by examining measures of liquidity, solvency, and profitability, but no single measure is entirely adequate to indicate a firm's vulnerability to financial stress. Rather, a combination of measures is generally desirable (Johnson et al. 1987; Lines and Morehart 1986).

Specific Measures of Financial Stress

Events of the past few years have stimulated considerable interest in the financial status of U.S. farms and ranches, and numerous analyses of farm financial conditions have resulted (see, for example, Johnson et al. 1986; Joseph and Reinsel 1986; Barry 1986; Harrington and Stam 1985; Lee 1986; Lines and Zulauf 1985; Salant et al. 1986; Watt et al. 1986). These analyses have used a number of measures of financial stress, and some of these measures are discussed in this section.

Ability to Make Debt Payments

One commonly used measure of financial stress is whether the farmer is current on loan payments. This simple and seemingly unambiguous indicator has been used in a number of farm finance analyses (Watt et al. 1986; Leholm et al. 1985; Murdock et al. 1985; Fiske et al. 1986; Mortensen 1987). However, as a measure of financial stress, it is not totally satisfactory because it does not indicate the firm's potential to adjust to the situation. For example, a farm could be profitable and have a strong equity position but encounter cash flow problems because most of its liabilities are short-term with a heavy schedule of principal payments. In such circumstances, the operator could probably resolve his difficulties by rescheduling his loan over a longer time period (Pederson et al. 1987). On the other hand, an operator whose inability to make scheduled payments stems from unprofitable enterprises and who has very little equity would probably not be able to resolve his problems through debt restructuring. Rather, a partial liquidation of assets and reorganization of farm enterprises might be the only option offering much hope of success (Brake and Boehlje 1985).

Net Cash Flow

A second type of indicator of financial stress, which has often been used either singly or in combination with other measures, is the farm's annual net income or net cash flow. A number of variations of the measure have been used, and these differ primarily because of variations in the types of income sources and costs that are included (see, for example, Salant et al. 1986; Lines and Morehart 1986; Leistritz et al. 1985b; Jolly et al. 1985a; Joseph and Reinsel 1986). In turn, the decision of which income sources and costs to include may depend in large measure on the analyst's view of the relevant time frame for analysis. A short-term view of farm financial problems, for instance, would suggest that all income sources of a farm household (including off-farm income) be included and that only current cash outlays be considered. On the other hand, a longer-range view would suggest that such noncash costs as depreciation be included because these costs cannot be deferred indefinitely (Lines and Morehart 1986; Leistritz et al. 1987e). A long-range view also casts doubt on the inclusion of income from nonfarm sources—a farm household may be willing to use such income to subsidize farm losses for a short time but not indefinitely.

Indicators of cash flow or income adequacy are often expressed in terms of the number of dollars by which the sum of cash inflows exceeds (or falls short of) the sum of demands for cash outlays. A shortcoming of this approach is that no provision is made to relate the amount of the cash-flow surplus or deficit to the total magnitude of cash inflow or outflow. Thus, it cannot be readily determined whether a 5 percent increase in cash inflow would resolve the problem or whether an increase of 50 percent would be required. In an effort to solve this problem, Salant et al. (1986) introduced the *viability ratio* as a measure of the financial well-being of farm households.

Viability Ratio

The viability ratio is defined as annual household net income divided by annual household financial obligations. The numerator of the ratio includes net cash income from the farm operation plus off-farm earned and unearned income. The denominator of the ratio is the sum of minimum household consumption (family living expenses), principal payments on debt obligations, and capital replacement costs. A ratio of 1.0 or greater indicates that the farm household is able to improve its well-being during the year. Farm households with a ratio of less than 1.0 are unable to meet all farm and household financial obligations. Some advantages of the viability ratio, compared to such measures as net farm income and net cash flow, are that it enables comparisons among varying farm sizes, farm types, and geographic regions and that it accounts for financial well-being from a farm household and not just a farm business perspective (Salant et al. 1986). On the other hand, this measure can be criticized because off-farm income can conceal the financially unhealthy state of the farm business and because some double-counting of financial obligations occurs when both depreciation and principal payments are incorporated.

Other indicators of current earnings or income of farm households, which have sometimes been used by analysts, include net cash farm income (Johnson et al. 1986; Leholm et al. 1985), net operating margin (Joseph and Reinsel 1986), total family income (Joseph and Reinsel 1986; Leistritz et al. 1985b; Johnson et al. 1986b), and return on equity (Johnson et al. 1987).

Leverage Position

A third type of indicator of financial stress is derived from the balance sheet and reflects the relationship between total assets and liabilities. The firm's overall debt-to-asset ratio is an indication of its solvency and risk-bearing ability. In addition, during periods of high real interest rates and relatively low returns to farm assets, highly leveraged firms are quite likely to experience cash-flow problems because the rate of return to assets is often less than the interest rate that must be paid. As a result, return to equity (the owner's own capital) may be near zero or even negative.

An example can help clarify this relationship. Table 4.2 illustrates the position of farm operators under three different leverage situations

Table 4.2

**Return-to-Owner Equity Under Alternative Leverage
Positions and Levels of Return to Assets and Interest Rates**

	Leverage Position		
Scenario	No Debt	Debt-to-Asset Ratio=0.4	Debt-to-Asset Ratio=0.7
Total assets	$300,000	$300,000	$300,000
Total debt	0	$120,000	$210,000
Owner equity	$300,000	$180,000	$90,000
Scenario I			
Return to assets (@ 10%):			
Dollars	$30,000	$30,000	$30,000
Percent	10.0	10.0	10.0
Interest paid (@ 6%)	0	$7,200	$12,600
Return to equity:			
Dollars	$30,000	$22,800	$17,400
Percent	10.0	12.7	19.3
Scenario II			
Return to assets (@ 6%):			
Dollars	$18,000	$18,000	$18,000
Percent	6	6	6
Interest paid (@ 10%)	0	$12,000	$21,000
Return to equity:			
Dollars	$18,000	$6,000	$-3,000
Percent	6.0	3.3	-3.3

Table 4.3

**Average Rate of Return to Total Assets and to Owner
Equity, North Dakota Farm Operators, 1985**

Debt-to-Asset Ratio	Return to Total Assets[1]	Return to Owner Equity[2]
No debt	5.1	4.4
.01-.40	3.6	0.3
.41-.70	5.4	-2.9
> .70	8.8	-25.0
Total	4.9	-1.3

[1](Net cash farm income plus interest paid minus family labor
allowance) divided by total farm assets.
[2](Net cash farm income minus family labor allowance) divided by
owner equity.

Source: Leistritz et al. 1987e.

when confronted with two contrasting levels of returns to assets and interest rates. The operator is assumed to have total farm assets of $300,000 and alternative leverage positions of no debt, 40 percent debt, and 70 percent debt. The return to assets, which is the total farm income less farm operation expenses and an allowance for operator and family labor, is equal to 10 percent of the total assets (i.e., $30,000) in the upper portion of the table (Scenario 1). This is the amount available to service debt and to provide a return to owner equity. The interest rate paid is assumed to be 6 percent. The return to equity, which is the return to assets less interest payments, will be the same as the return to assets for the operator with no debt. For the operator with a debt-to-asset ratio of 0.4, the return to equity will be $22,800 ($30,000 less $7,200) or a 12.7 percent return on owner equity. Return to equity exceeds return to assets because the operator is essentially earning a profit of 4 percent on the borrowed funds. When this type of relationship between return to assets and cost of borrowed funds is expected to prevail, expansion through use of debt financing is attractive.

The lower portion of Table 4.2 illustrates the situation when returns to assets are less than interest rates (Scenario 2). Suppose the rate of return to farm assets has fallen to 6 percent while interest rates have risen to 10 percent. The operator with a debt-to-asset ratio of 0.4 must pay $12,000 in interest out of his total return to assets of $18,000. His remaining return to equity is $6,000 (or only 3.3 percent), and depending on his schedule of principal repayment, this amount may not be adequate to meet the scheduled payments. The operator with a debt-to-asset ratio of 0.7, however, has an even more serious problem as his total return to assets is not adequate even to cover interest payments.

This latter example appears to be typical of the situation confronting many highly leveraged producers during the last few years. The return on assets and the return on equity experienced by a group of 759 North Dakota farm and ranch operators during 1985 are summarized in Table 4.3. These operators, who are believed to be representative of North Dakota producers who are less than age 65 and consider farming to be their primary occupation (Leistritz et al. 1987e), averaged a 4.9 percent return to assets and a negative 1.3 percent return to equity. Note, however, that return to equity averaged 4.4 percent for operators with no debt and 0.3 percent for those with debt-to-asset ratios of 0.4 or less. Producers with debt loads of over 40 percent experienced negative returns to equity and those with debt loads over 70 percent had a return to equity of negative 25 percent, indicating a serious erosion of their equity.

Solvency and Earnings

Because both a farm's current earnings and its solvency position are important indicators of its financial vulnerability, some analysts have suggested combining earnings and solvency measures. For example, Jolly et al. (1985) used a combination of net cash flow and debt-to-asset ratio in identifying the incidence and intensity of financial stress. Net cash flow was defined as including income over farm cash expenses plus off-farm income less withdrawals for consumption, taxes, and debt service. Their analysis of data from a nationwide survey indicated that about 19 percent of all farm operators had debt-to-asset ratios over 40 percent and that about one-half of all farmers had negative net cash flows in 1984. If a joint criterion is used, 12.6 percent of U.S. farmers had debt-to-asset ratios greater than 0.4 and a negative net cash flow; these operators had both solvency and liquidity problems. On the other hand, about 43 percent of U.S. farmers had debt ratios less than 0.4 and produced a positive cash flow.

Several combinations of solvency and earnings measures were used by Johnson et al. (1987). These analysts not only used a combination of cash-flow and debt-to-asset-ratio criteria similar to that employed by Jolly et al. (1985) but also incorporated two additional approaches in addressing the financial well-being of commercial farmers. The first of these, adopted from earlier work by Melichar (1986), uses measures of return to assets, return to equity, debt-to-asset ratio, and dollar amount of owner equity to evaluate financial vulnerability. The second incorporates the farmer's debt-to-asset ratio and a debt-service ratio, which measures the adequacy of income to service debt. The debt-service ratio is intended to measure the income available to pay principal and interest payments. It is the sum of net cash income before interest (including off-farm income) minus machinery investment and estimated household withdrawals divided by the sum of principal and interest expenses. A debt-service ratio of one or more implies full debt service, but a ratio between zero and one suggests that only partial debt service can occur. A ratio of zero or less indicates no debt service ability if the other business and household expenses are met.

In summary, measures of financial stress are generally derived from one of the three basic financial statements: the balance sheet, the income statement, or the cash flow statement. These indicators are designed to measure one or more of three financial characteristics: solvency, liquidity, or profitability. Differences in the specific measures employed in assessing financial stress often arise from alternative views concerning the relevant time frame for analysis and whether the farm household or the farm business is the appropriate unit to examine. These, then, are some measures that have been used to estimate the incidence and intensity of financial stress in U.S. agriculture. In the following section, some of these measures are employed in describing the financial status of farmers and ranchers located in different regions of the country and operating farms of different types and sizes.

Current Extent of Farm Financial Vulnerability

A number of analyses of the financial status of U.S. farmers have been conducted over the last few years. These have ranged from nationwide analyses (for example, Johnson et al. 1985; Johnson et al. 1986; Jolly et al. 1985a) to assessments of conditions in a given state (Leholm et al. 1985; Hardesty 1986; Watt et al. 1986; Jolly and Barkema 1985) and to evaluations specific to selected substate regions (Salant et al. 1986). Thus, a number of data bases could potentially be used in defining the magnitude and distribution of financial stress in U.S. agriculture. Because a major purpose of this work is to provide a comprehensive view of farm financial problems on a nationwide basis, the approach taken here will be first to describe the extent of financial stress across farming regions and farm types using data bases that provide comparable information on a nationwide basis. Then, attention will be turned to selected state-specific analyses that provide insights into selected aspects of the problems confronting farmers and ranchers.

Farm Costs and Returns Survey

The most comprehensive data base concerning the financial status of U.S. farmers and ranchers is provided by the Farm Costs and Returns Survey (FCRS), which is conducted annually by the National Agricultural Statistics Service, U.S. Department of Agriculture. This survey provides information regarding both farm earnings and solvency, as well as data concerning production practices and costs for selected enterprises. The most recent survey from which data were available at the time of this writing (mid-1987) is the survey covering the 1985 production year. Conducted in February and March of 1986, the survey yielded 11,497 sample observations. The survey is generally regarded as being representative of farms and ranches across the country except that farms with less than $10,000 in sales tend to be undercounted (Johnson et al. 1986). Most of these farms would generally be regarded as part-time or hobby farms.

Solvency and Cash Flow. Data from the FCRS provide a basis for assessing farm financial vulnerability. Table 4.4 provides a summary of solvency and cash flow information for U.S. farms and ranches by farm production region.

Examination of Table 4.4 indicates that considerable variation in the debt-to-asset position exists among regions (Figure 4.1). Nationwide, 12.7 percent of farmers had debt-to-asset ratios between 0.4 and 0.7, 4.6 percent had ratios of 0.71 to 1.0, and 4.0 percent had ratios greater than 1.0. Analysts have noted that, under conditions prevailing in recent years, farms producing major agricultural commodities often begin experiencing difficulty meeting all of their financial obligations at a debt-to-asset ratio of about 0.4, while at a level of about 0.7 they often find they cannot even meet interest payments (Johnson et al. 1985). Farms with debt-to-asset ratios greater than 1.0 are, of course, technically insolvent. Considering the information in Table 4.4, then, it appears that slightly over one in five U.S. farms have debt-to-asset ratios that are typically associated with financial vulnerability.

Another financial indicator that is relevant in determining potential vulnerability is the earnings or cash flow of the farm business or household. Analysis of the 1985 FCRS data revealed that about 48 percent of all farm businesses surveyed had a negative net cash income in 1985. The farm business net cash income includes all revenue from farm-related activities less all cash expenses including interest paid but excluding principal payments (Johnson et al. 1985). If other farm household income and outlays were considered, about 45 percent of all the farms surveyed had a negative net cash income in 1985. (Farm household net cash income includes all sources of cash income including nonfarm earnings minus all cash expenses including interest plus an estimate of principal payments and a family living allowance.)

One aspect of the relationship between earnings and solvency for U.S. farms is summarized in Figure 4.2. This figure shows what proportions of farms had negative and positive levels of farm household net cash income and what proportions had high (i.e., greater than 0.4) and low (i.e., less than 0.4) debt-to-asset ratios. The figure shows that 11.2 percent of all farms combined a high debt-to-asset ratio with a negative cash flow. These producers could be seen as the most vulnerable to financial pressures. On the other hand, about 45 percent had both low debt-to-asset ratios and positive cash flow and could be regarded as occupying a relatively strong financial position. Examination of Figure 4.2 also shows that, while there appears to be some relationship between high debt ratios and negative cash flow, this relationship is not pronounced. Of farms with high debt ratios, 52.6 percent had negative cash flow in 1985, compared to 42.6 percent of those with low debt ratios.

Table 4.4

Distribution of Farms with High Debt-to-Asset Ratios and Negative Cash Flows, by Farm Production Region, 1985

Region[1]	Debt-to-Asset Ratio[2]				High Debt-to-Asset Ratio and Negative Cash Flow[3]
	.40-.70	.71-1.00	> 1.0	Total, .40 or more	
	(- - - percent of all farms in region - - -)				
Northeast	9.3	3.3	1.4	14.0	6.6
Appalachia	6.7	1.1	6.4	9.3	5.7
Southeast	9.8	3.4	5.1	15.8	7.9
Delta	7.7	3.0	6.8	16.5	11.3
Corn Belt	15.6	5.6	1.5	26.3	11.7
Lake States	19.1	7.3	2.6	32.8	19.8
Northern Plains	17.6	8.8	5.8	33.2	17.1
Southern Plains	9.0	3.2	3.0	15.2	8.0
Mountain	16.0	4.9	2.9	23.8	12.2
Pacific	10.5	4.0	2.1	16.6	7.8
All	12.7	4.6	4.0	21.3	11.2

[1]See Figure 4.1.
[2]Debt-to-asset ratio as of December 31, 1985.
[3]Cash flow refers to net cash operating income of farm households and reflects estimated principal payments, nonfarm income, and family living allowances.

Source: Petrulis et al. (1987) and Johnson et al. (1986).

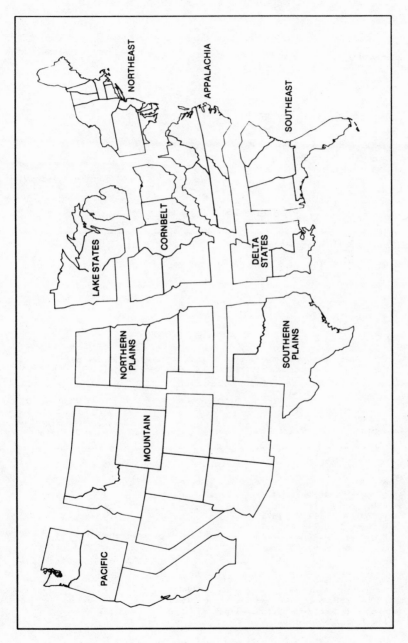

Figure 4.1 Farm production regions.

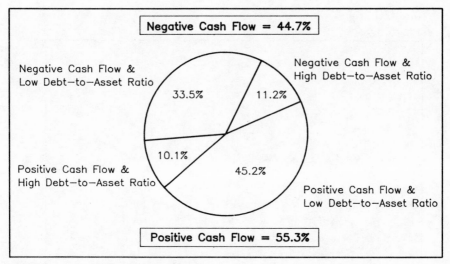

SOURCE: Johnson, Baum, and Prescott 1985

Figure 4.2 Cash flow and debt—to—asset ratios of U.S. farmers, 1985.

Table 4.5

**Distribution of Farms with High Debt-to-Asset Ratios and
Negative Cash Flows, by Sales Class, 1985**

Sales Class	Debt-to-Asset Ratio[1]				High Debt-to-Asset Ratio and Negative Cash Flow[2]
	.40-.70	.71-1.00	> 1.0	Total, .40 or more	
	(- - - - percent of all farms in class - - - -)				
$500,000 or more	21.9	10.1	7.9	39.9	14.0
$250,000 to 499,999	25.3	10.4	6.1	41.8	14.8
$100,000 to 249,999	20.4	9.1	7.9	37.4	17.8
$40,000 to 99,999	16.4	5.6	5.9	27.9	17.6
$20,000 to 39,999	11.3	4.2	3.3	18.8	11.8
$10,000 to 19,999	9.2	2.7	4.9	16.8	10.8
Less than $10,000	7.1	2.0	0.9	10.0	4.4
All	12.7	4.6	4.0	21.3	11.1

[1]Debt-to-asset ratio as of December 31, 1985.
[2]Cash flow refers to net cash operating income of farm households
and reflects estimated principal payments, nonfarm income, and
family living allowances.

Source: Petrulis et al. (1987) and Johnson et al. (1986).

Returning to Table 4.4, it is now possible to combine information concerning debt position and cash flow to identify the incidence of farm financial vulnerability. First, examination of the percentage of farms in each region that have debt-to-asset ratios greater than 0.4 suggests substantial regional disparities. This group makes up 21.3 percent of farms nationwide, but the percentage ranges from highs of 33.2 percent in the Northern Plains, 32.8 percent in the Lake States, and 26.3 percent in the Corn Belt to lows of 9.3 percent in Appalachia and 14.0 percent in the Northeast. When cash flow information is combined with debt position, about 11.2 percent of all farmers were found to have both high debt and negative cash flow. This percentage again shows substantial variation, ranging from 19.8 percent in the Lake States and 17.1 percent in the Northern Plains to 5.7 percent in Appalachia.

Farm Sales Class. The same measures of financial stress can be used to examine the incidence of farm financial stress by sales class of farm (Table 4.5). When the percentage of farms with debt-to-asset ratios in excess of .4 is examined, it is apparent that the proportion of highly leveraged farms is greater in the higher sales classes. However, when cash flow information is considered together with the leverage situation, it becomes clear that the percentage of farms with both high debt ratios and negative cash flow is greater in the sales classes of $40,000 to $99,999 and $100,000 to $249,999. Thus, the sales classes that many would describe as family-size commercial farms appear to be most vulnerable to financial stress.

Farms with sales of less than $40,000 are less likely to have incurred large debts for farm acquisition or expansion, and their operators rely primarily on nonfarm income for their livelihood. Farms with sales exceeding $250,000, on the other hand, often are engaged in enterprises with relatively high returns, such as speciality crops or cattle feeding, or are able to achieve substantial cost advantages per unit of output, either through greater efficiency in production or through input purchasing and marketing advantages (for example, see Smith et al. 1984). The moderate-size commercial farms depend heavily on farm income for family livelihood but are unable to fully capture the cost advantages of a large-scale operation.

Enterprise Type and Operator Age. The composite criterion, which considers both debt load and cash flow, also can be used to examine the incidence of financial stress by farm enterprise type and age of operator (Table 4.6). The percentage of financially vulnerable farms is above average for dairy and cash grain farms. Dairy producers have recently been confronted with the problem of adjusting to substantial production surpluses and depressed prices, while cash grain producers have been substantially impacted by declining export markets. The

Table 4.6

**Distribution of Farms with High Debt-to-Asset Ratios and
Negative Cash Flows, by Farm Type and Operator Age, 1985**

Farm Type/ Operator Age	Percentage with High[1] Debt-to-Asset Ratios and Negative Cash Flow[2]
Farm Type	
Dairy	20.2
Cash grain	13.3
General crop	10.5
Field crop	10.5
Vegetable, fruit, nut	9.8
General livestock	8.3
Poultry	8.0
Other livestock	7.8
Nursery and greenhouse	2.1
All types	11.1
Age of Operator	
Less than 35	22.3
35 to 44	17.3
45 to 54	10.9
55 to 64	6.6
65 and over	7.2
All ages	11.1

[1] Debt-to-asset ratio as of December 31, 1985.
[2] Cash flow refers to net cash operating income of farm households
and reflects estimated principal payments, nonfarm income, and
family living allowances.

Source: Petrulis et al. 1987.

percentage of vulnerable farms also is greater among producers in the younger age groups. It appears likely that the operators less than age 44 largely entered farming during the 1970s and thus were engaged in acquiring land and other assets during a period when values were inflated. They then entered the 1980s with substantial debt loads and only limited equity in their farming operation. As noted earlier, such operators tend to be particularly vulnerable during periods of depressed returns, high interest rates, and falling asset values.

Summary. The analysis of data from the FCRS survey covering the 1985 production year provides a snapshot of the financial position of U.S. farm operators at one point in time. The analysis indicates that at the end of 1985 more than one in five producers had debt-to-asset ratios in the range often associated with financial vulnerability and that between 45 and 48 percent (depending on whether off-farm income and principal repayment obligations are considered) experienced negative income or cash flow in 1985. About 11 percent of all producers had both high debt ratios and negative cash flow; these operators can be considered highly vulnerable, and the continuation of their farming operations may well be in jeopardy. Geographically, these farmers are most frequently found in the Lake States and the Northern Plains where more than one operator in six has these attributes. These highly vulnerable farms are most frequently family-size commercial farms with annual sales in the range of $40,000 to $249,999. The vulnerable farms are found most frequently among those specializing in dairy or cash grain production, and their operators tend to be in the younger age groups (i.e., less than 45).

Other Analyses

Other analyses add additional perspectives concerning the characteristics of farms that are financially vulnerable. In particular, they provide insights concerning the financial status of commercial farms compared to smaller ones, of farmers with different proportions of rented land, and of farms specializing in different enterprises.

Farm Sales. A survey of Michigan farmers indicates that farms with gross sales exceeding $100,000 were in somewhat worse financial condition than smaller farms (Hardesty 1986). Although about 48 percent of the farms with sales over $100,000 had debt-to-asset ratios exceeding 0.4, only 16 percent of the smaller farms fell into this category. The larger farms also were more likely to be delinquent on debt payments. However, the farms with sales over $100,000 (which accounted for only 22 percent of the farms but 78 percent of the total sales in the sample) also earned substantially more profit than the smaller farms. These farms had an average net cash farm income of $30,283 in 1985 compared to only $1,557 for the smaller operations. The author concludes that larger farms that are not highly leveraged earn substantial net incomes from farming, but those that are highly leveraged are not able to earn enough income to meet their debt servicing commitments (Hardesty 1986).

While smaller farms are less likely to be highly leveraged, they are also less likely to generate adequate earnings to cover all costs of farm operation including depreciation. In an analysis that focused on the adequacy of farm income and included depreciation as a cost while excluding off-farm income, Lines and Morehart (1986) concluded that the proportion of farms with sales less than $40,000 that could not cover all costs was greater than the proportion of larger farms. These authors question whether many small farms can be viable in the long term if their operators do not continue to use off-farm income to subsidize the farming operation.

Rented Land. The evidence concerning the effect of the percentage of rented land on financial stress is similarly mixed. In a survey of Iowa producers, Jolly and Olson (1986) concluded that farmers in the most vulnerable category (i.e., highly leveraged and with inadequate earnings) also had the highest proportion of rented land. Similarly, analyses of survey data from Texas (Murdock et al. 1985) and Ohio (Lines and Zulauf 1985) indicate that operators who rent a large portion of their land are more likely to be highly leveraged. On the other hand, a nationwide analysis that used the adequacy of farm income to cover all costs of operation as its criterion concluded that farms with a higher proportion of rented land were more likely to experience good financial health (Lines and Morehart 1986).

Enterprise Type. Analyses of the financial status of farms specialized in different enterprises also yield varying conclusions. A survey of Ohio farmers found that specialization in grain production did not significantly affect a farm's debt-to-asset ratio, compared to farms specialized in dairy or other livestock or those diversified between grain and livestock (Lines and Zulauf 1985). On the other hand, a study of Texas producers found that crop farms were most likely to be highly leveraged (Murdock et al. 1985), and a North Dakota analysis indicated that dairy and beef producers averaged substantially lower levels of net cash farm income than grain producers in both 1984 and 1985 (Leistritz et al. 1985b, 1987e). Finally, in analyzing the 1985 FCRS data using a combination of returns, equity, and solvency criteria, Johnson et al. (1987) found that producers of cotton, rice, and beef were most likely to be classified as financially vulnerable. Perhaps the safest conclusion that can be drawn is that analyses of financial status by enterprise type are likely to yield different conclusions depending on the region studied and the criteria employed. Further, analyses based on earnings may yield differing conclusions depending on the year selected for study because of the effect of fluctuating market conditions on the returns received from different enterprises.

The information from the FCRS survey, together with that from other analyses, provides a valuable overview of the farm financial situation, but some additional questions remain to be explored. Perhaps the most important of these concerns trends in the financial status of farmers and ranchers. Has their financial situation stabilized, as some suggest, or is it continuing to deteriorate? This question is addressed in the section that follows.

Trends in the Financial Status of Farm Operators

Evaluating trends in farm financial status requires evaluating changes in both farm earnings and solvency measures. Because it is possible for these measures to move in different directions, changes in overall farm financial health or vulnerability can be difficult to discern. Further, the evaluation of such trends is complicated by the fragmentary nature of available data sources. Data collected from farm surveys often are not completely comparable between survey years, and the most recent data available generally are for the 1985 production year.

Farm Earnings

Aggregate data for the farm sector indicate that returns to assets improved in the 1985–1987 period compared to the 1980–1984 period (Johnson et al. 1987). Gross farm income for the period 1985–1987 ranged from $127.3 billion (in 1982 dollars) to $139.9 billion, compared to an average of $152.3 billion for the 1980–1984 period, but the return to assets for the more recent time period ranged from $20.2 billion to $25.0 billion, compared to an average of $18.9 billion for the 1980–1984 period (Table 4.7). Returns to equity also were higher, ranging from $6.7 to $9.1 billion, compared to a negative $0.8 billion in 1980–1984. When expressed in percentage terms, the returns to assets and equity show even more improvement, but the major reason is the substantial decreases in asset and equity values that have occurred in recent years. The erosion of asset and equity values is anticipated to continue at least through 1987 (Table 4.7), and consequently, the debt-to-asset ratio for the sector as a whole will continue to rise.

State-level surveys tend to support the findings of this national analysis. For example, statewide surveys conducted in North Dakota and Texas indicated that net cash farm income was higher in 1985 than in 1984

Table 4.7

Returns to U.S. Farm Production Assets and Equity, 1980-1987

Farm Financial Characteristics	Year			
	1980-84	1985	1986[a]	1987[a]
	(- - billions of 1982 dollars - -)			
Gross farm income (excluding operator households)	152.3	139.9	131.4	127.3
Return to farm assets	18.9	25.0	20.2	21.8
Return to equity	-0.8	8.9	6.7	9.1
Total farm assets (Dec. 31 of previous year)	994.1	766.4	673.7	597.9
Total farm equity (Dec. 31 of previous year)	807.2	588.5	505.9	441.1
	(- - - - - percent - - - - -)			
Return to farm assets	2.0	3.3	3.0	3.7
Return to farm equity	-0.1	1.5	1.3	2.1

[a]Projected values.

Source: Johnson et al. 1987.

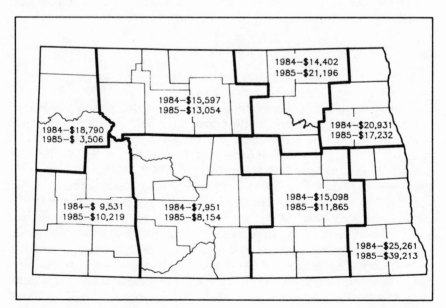

SOURCE: Leistritz et al. 1987c

Figure 4.3 Regional average net cash farm income in North Dakota, 1984 and 1985.

(Leistritz et al. 1987e; Albrecht et al. 1987a). However, the North Dakota analysis also indicated a high degree of income variability within the state, based on regional variations in weather conditions and differences in market conditions for various commodities. As shown in Figure 4.3, regional average net cash farm income in North Dakota in 1985 ranged from $3,506 in the northwest region (which suffered a severe drought) to $39,213 in the southeast. The changes in net cash farm income at the regional level from 1984 to 1985 were also quite substantial although the change was only 4.4 percent at the state level.

Solvency

State-level analyses also indicate the general trend of decreasing farm operator equity and indicate that declining asset values are the primary cause. For example, in Texas the average per-farm value of assets fell from $815,000 as of January 1, 1985, to $641,000 one year later, primarily as a result of declining land values, but the average level of debt remained virtually unchanged (Albrecht et al. 1987a). Similarly, in North Dakota average asset value dropped 3.7 percent from January 1, 1985, to January 1, 1986, but debt increased by slightly less than 1 percent (Leistritz et al. 1987e).

The state surveys also point out the stark plight of many highly leveraged operators. For example, Iowa farmers with debt-to-asset ratios in the 0.7 to 1.0 range lost 88 percent of their equity between 1984 and 1985—falling asset values swamped their efforts to reduce their debt (Jolly and Barkema 1985). Similarly, North Dakota farmers with debt-to-asset ratios between 0.7 and 1.0 experienced a return to equity of negative 25 percent in 1985, indicating that one-fourth of their equity was lost simply through the inability of their farming revenues to cover all costs (Leistritz et al. 1987e).

The North Dakota analysis also suggests that the plight of the highly leveraged operator is likely to be chronic. Of a panel of 759 farm operators who provided financial data for both 1984 and 1985, 42 percent had insufficient total family income to cover all their financial obligations (including farm operation costs, depreciation, interest, principal repayment, and a family living allowance) in both of those years, and 64 percent had incomes inadequate to cover all expenses in at least one year. The 42 percent whose income was inadequate in both years had an average debt-to-asset ratio of 0.55, but those whose income was adequate in both years had an average debt-to-asset ratio of 0.15 (Leistritz et al. 1987e).

Summary

The financial status of many U.S. farmers continues to erode as falling land values lead to declining equity and push many to the brink of insolvency. Improved returns to assets since 1984 coupled with some reductions in real interest rates should provide strength to the land market, and, although land values may continue to decline for at least another year or two, the decrease should not be as precipitous as during 1984–1985 (Harrington and Carlin 1987; Johnson et al. 1987). Current levels of farm income and returns, however, apparently can be maintained (in the short term) only through continuation of current levels of federal farm program payments. Questions about the future of farm commodity programs, then, create a major uncertainty regarding future changes in the farm financial status. Barring major changes in world markets for agricultural commodities, in U.S. monetary and fiscal policy, and in U.S. farm programs, highly leveraged producers will continue to experience substantial financial stress. How they attempt to cope with the situation will likely have long-term effects on the structure of agriculture and on many rural communities.

Conclusions and Implications

The purpose of this chapter was threefold: (1) to define the meaning of financial stress; (2) to examine alternative measures of actual or potential financial stress; and (3) to examine the extent of financial stress among farms in different regions of the country, among farms of different sizes, and among farms producing different types of commodities. Financial stress can be defined as resulting from the inability to meet financial obligations when due, and a realistic assessment of the extent of actual or potential financial stress requires examining both earnings (or cash flow) and solvency characteristics of farming operations. Specific issues in measuring stress include whether to include depreciation and off-farm income in evaluating the adequacy of earnings to meet financial obligations and the relative weight to be given to earnings and solvency.

Analysis of data from the nationwide Farm Costs and Returns Survey for the 1985 production year provides insights concerning the characteristics of producers experiencing financial stress. Highly leveraged farms (with debt-to-asset ratios exceeding 0.4) were found to be concentrated in the Lake States, the Northern Plains, and the Corn Belt. Farms with both high debt-to-asset ratios and negative cash flow were concentrated especially in the Lake States and Northern Plains. These financially vulnerable farms were also found to be most prevalent among farming operations with annual sales from $40,000 to $249,999 and among those specializing in dairy and cash grain production. These appear to be most vulnerable to an early forced exit from agriculture, yet other analyses indicate that many farms with annual sales less than $40,000 will be able to continue only as long as their operators are willing to use income from nonfarm sources to subsidize the farming operation.

Recent trends indicate that farm sector earnings since 1985 have been somewhat better than those experienced during the years 1980–1984 but that the sector's asset values and equity continue to deteriorate, primarily because land values have continued to decline. Many highly leveraged farm operators continue to find their income inadequate to meet all of their financial obligations, and they are also extremely vulnerable to substantial deterioration of their equity if land values continue to fall. These producers will continue to face harsh, and in many cases insurmountable, challenges in the years ahead.

5

Producer Reactions and Adaptations

F. Larry Leistritz, Brenda L. Ekstrom, Harvey G. Vreugdenhil and Janet Wanzek

Financial pressures resulting from changing economic conditions have placed some farm and ranch families in positions of considerable financial stress. These families have an obvious incentive to examine whether possible changes in their farm management practices, farm organization, farm financial structure, or family lifestyle could enable them to meet their financial obligations. If such adjustments still will not enable income to cover expenses, the farm household may be forced either to rely more heavily on off-farm income sources or to liquidate the farm and seek alternative employment. This chapter examines a number of changes in farm management practices, farm and financial organization, and family lifestyles initiated by farm and ranch operators to cope with financial stress. Farm liquidation as another option in responding to conditions of financial stress is also discussed in this chapter, while off-farm employment is treated in chapter 6.

Nature of Alternative Adjustment Strategies

The theoretical basis for the sequence of adjustments that a financially stressed farm household may undertake is not well developed even though firms in virtually all industries face a similar array of possible adjustments. This section discusses adjustment measures that (1) increase or stabilize net income, (2) restructure liabilities, or (3) restructure assets (Brake and Boehlje 1985). Should these measures prove inadequate to solve the firm's problems, a final alternative is to liquidate its remaining assets. (In some industries where larger firms predominate, an equity infusion, merger, or acquisition by another firm may be an alternative to liquidation, but these options are seldom used by the typical farm.)

Measures to increase or stabilize net income include those aimed at increasing the value of sales, reducing costs, and stabilizing the farm's net income over time by reducing the risk of experiencing a particularly unfavorable year. Strategies to increase the value of farm sales include adding new enterprises (such as incorporating a livestock enterprise into a crop farm), more intensively managing existing enterprises (for instance, better management of chemicals and fertilizer for crops or closer attention to disease prevention in livestock), and increasing attention to marketing the farm's products advantageously. Attempts to reduce costs can include reducing the use of selected inputs (e.g., fertilizer and chemicals) or reducing the prices paid for resources (e.g., renegotiating a land rental contract). Periods of financial stress also may encourage strategies that minimize immediate outlays even though the associated costs may be higher in the longer run. For instance, a farmer may postpone capital purchases or may choose to lease machinery rather than purchase it in an effort to solve short-run cash flow problems.

Efforts to stabilize incomes and reduce the probability of particularly unfavorable income years include a wide range of production and marketing strategies (Boehlje and Eidman 1983). For instance, shifting land rental contracts from a cash to a crop-share basis shifts some of the risk of a bad crop to the landlord. The use of crop insurance can often provide some protection against crop failure, and the use of contracting or hedging as marketing tools can reduce the risk associated with fluctuating commodity prices. Participation in government farm programs also can be used as a risk reducing strategy by some producers because the major commodity programs generally offer some income protection through deficiency payments, nonrecourse loans, diversion payments, and other program provisions.

In addition to attempts to increase or stabilize net farm income, some farm households may attempt to reduce family living expenses. These efforts may represent a reduction in all categories of household costs or may be focused on certain classes of expenditures that often seem easier to defer.

Because many farmers have been unable to increase income and cash flow sufficiently to meet all their obligations, many have attempted to restructure their liabilities in order to reduce debt service requirements. Restructuring of liabilities can take a variety of forms including deferral of interest and/or principal payments, reamortization of loan principal over longer repayment periods, or even renegotiating the loan to reduce the interest rate, the loan principal, or both.

When restructuring liabilities proves inadequate to relieve financial stress, many farms must consider restructuring assets. Asset restructuring is a euphemistic term that generally implies liquidation of some of the farm's assets. Assets with a relatively long useful life, and typically a low annual yield, are those most often targeted for liquidation. The proceeds from the sale of the assets would normally be applied to reduce debt obligations. The success of such a strategy, however, depends on the ability to obtain a favorable price for the liquidated assets and on the firm's ability to continue efficient operations after their disposal. Sale-leaseback arrangements where ownership rights are transferred but use rights to the assets are retained are a means of addressing the latter issue, but the depressed markets for farm real estate and machinery, which have characterized the past few years, have seriously limited the potential for many farmers to successfully restructure their assets (Doye et al. 1987).

If asset restructuring proves unsuccessful as a means of maintaining a viable farming unit, liquidation of the farm business may be the only alternative. Choices at this stage include different means of asset disposition. For example, proceeds from the sale of land, machinery, and other major assets may be used to repay debts, or such assets may be voluntarily conveyed to creditors, who hold security interests in the property, as a partial or complete satisfaction of their claims. In some cases, creditors may initiate foreclosure proceedings to satisfy their claims, and some farmers elect bankruptcy to obtain protection from their creditors. On the other hand, when the farm business being liquidated is still relatively solvent, the operator may be able to retain some assets. For instance, the land might be retained and rented to another operator, while the machinery and livestock are sold. Finally, if the farming operation is liquidated, household members often must seek alternative employment and in some cases may chose to relocate to another area.

These, then, are some of the potential adaptations that farm households may make in response to conditions of financial stress. In the next section, the extent to which farm households have actually employed these alternative adjustment strategies is examined.

Use of Adaptation Strategies

Information concerning the extent to which various adjustment strategies are being employed by farm households is drawn from surveys conducted recently in several states. The surveys differed somewhat in structure as well as in sample design and administrative procedures, so the reader is cautioned that comparability of results among states is limited. Nevertheless, the survey results provide an indication of the extent to which various response strategies are being used by farm operators and their families.

Measures to Increase Net Income or Reduce Risk

A survey of North Dakota farmers and ranchers conducted in the spring of 1986 provides insight concerning the frequency with which producers are using alternative strategies to increase net income or reduce risk (Leistritz et al. 1987e). More than 62 percent of the producers surveyed indicated that they had postponed capital purchases during the past year, almost half of the respondents had reduced tillage, and a similar percentage had reduced family living expenses (Figure 5.1). More than 1 in 4 had reduced their expenditures for fertilizer, chemicals, and similar inputs, and 22 percent reported beginning or increasing their participation in government farm programs.

Farmers with heavy debt loads were more likely to report using such adjustment strategies. Table 5.1 summarizes the percentage of respondents in each of four debt-to-asset ratio categories who reported making selected management changes in 1985. Differences among debt levels were statistically significant for all but two changes.

Postponing capital expenditures and reducing family living expenses were also the adjustment strategies most frequently practiced by a group of Texas producers interviewed in 1986. Almost 73 percent had postponed capital purchases, and 65 percent had reduced living costs (Albrecht et al. 1987a). Other commonly reported adaptations included replacing machinery with used rather than new equipment (45 percent), cutting back on fertilizer and chemicals (43 percent), starting to participate in government programs (42 percent), and reducing tillage operations (42 percent). These authors also report that 26 percent of the respondents had increased the nonfarm uses of their land, such as hunting or oil and gas leases, as a means of responding to financial problems.

Producers in other parts of the country also report changing their management practices and lifestyles in response to changing conditions. More than 80 percent of Iowa farmers interviewed in 1987 indicated they were paying closer attention to marketing, while 32 percent had diversified by adding livestock, 31 percent had built on-farm grain storage, and 24 percent had started using forward contracting to buy farm inputs (Lasley 1987). Iowa farmers also reported efforts to reduce living costs, frequently indicating cutbacks in social activities and entertainment (72 percent), postponement of major household purchases (65 percent), changes in travel plans (57 percent), and changes in shopping or eating habits (56 percent). About 59 percent had to use savings to meet expenses (Lasley 1985).

A group of Ohio producers surveyed in 1985 most frequently reported participating in government programs, keeping more complete records, using contracting and hedging, and reducing living expenses as adap-

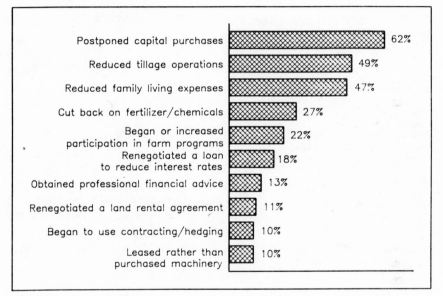

SOURCE: Leistritz et al. 1987e

Figure 5.1 Management adjustments to increase net income
or reduce risk, North Dakota farmers, 1985.

Table 5.1

**Management Changes Initiated by North Dakota Farm Operators
by Debt-to-Asset Ratio, 1985**

Management Change	Total	Debt-to-Asset Ratio			
		No Debt	.01-.40	.41-.70	>.70
	(- percent of operators reporting change -)				
Switched from cash to share rent	3.3	0	3.0	5.3	4.9
Renegotiated a land rental agreement to reduce land rents*	11.0	1.5	8.9	18.1	16.2
Began to use contracting or hedging as marketing tools	10.2	3.8	11.3	13.5	9.8
Began to use crop insurance*	10.7	3.0	10.0	11.1	20.3
Obtained professional financial advice*	12.9	1.5	10.9	14.6	27.6
Leased machinery rather than purchased*	10.3	4.5	8.3	14.0	16.3
Reduced family living expenses*	46.9	21.1	42.7	57.3	70.7
Postponed capital purchases*	62.3	33.8	56.0	79.5	84.6
Started participating in government farm programs*	9.6	6.0	9.9	6.4	17.0
Increased participation in farm programs*	12.1	3.8	12.6	12.9	18.7
Cut back on expenditures for fertilizer and chemicals*	26.7	15.8	26.5	24.0	42.6
Reduced tillage operations*	49.4	33.8	51.3	53.2	56.1

*Differences are statistically significant at the .01 level.

Source: Leistritz et al. 1987e.

tations they had already undertaken (Lines and Pelly 1985). These measures were also frequently reported as ones producers planned to undertake in the future, along with efforts to reduce debt, lower rental payments, and cut machinery costs.

Participation in government farm programs, maintaining more complete records, and developing financial budgets were the actions taken most frequently by Indiana producers interviewed in 1985 (Dobson et al. 1985). Many of these farmers also indicated that they had plans within the next year to reduce their debt, keep more complete records, participate in government programs, and reduce their machinery inventory.

While many producers have been attempting to raise their net incomes and reduce risk, others with relatively high debt loads have found that some form of debt or asset restructuring is a desirable, if not absolutely essential, adjustment to the changing economic climate. The next section takes a closer look at restructuring strategies undertaken by producers in North Dakota and Texas.

Debt and Asset Restructuring

Attempts to reduce indebtedness have been foremost in the minds of many farmers and agricultural lenders in recent years. Almost 18 percent of North Dakota farmers reported that they had renegotiated a loan to obtain lower interest rates during 1985, and more than 14 percent reported renegotiating a loan to reduce the principal amount (Figure 5.2). On the other hand, only about 1 percent had sold or deeded back land. All of the asset and debt restructuring measures shown in Figure 5.2 were significantly related to the farmer's debt-to-asset ratio except sale of land, for which the number of cases was insufficient for meaningful analysis.

A survey of Texas farmers also indicated that a substantial number had made efforts to restructure debt and/or assets (Albrecht et al. 1987a). Actions reported by 10 percent or more of these producers (not necessarily during the past year) included selling breeding livestock, renegotiating loans, changing lending institutions, and selling machinery (Table 5.2). All of these actions except selling land and breeding livestock were found to be significantly related to the respondent's debt load.

Liquidation of Farming Operation

Several recent studies provide insights concerning the characteristics of families who have been liquidating their farm businesses. A survey of 169 North Dakota producers who quit farming was conducted in 1986. Of the respondents, 79 percent had ceased farming in 1983, 1984,

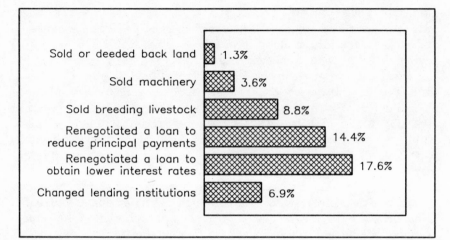

SOURCE: Leistritz et al. 1987e

Figure 5.2 Debt and asset restructuring measures undertaken
by North Dakota farmers, 1985

Table 5.2

Debt and Asset Restructuring Measures Undertaken by Texas Farmers

Restructuring Change	Total	Debt-to-Asset Ratio			
		No Debt	.01-.40	.41-.70	>.70
	(-percent of operators reporting change-)				
Sold land	8.3	9	10	20	13
Deeded back land*	1.9	2	1	5	6
Sold machinery*	11.7	6	13	26	33
Sold breeding livestock	24.5	29	29	23	28
Renegotiated a loan agreement or land contract to reduce the principal amount*	19.4	5	19	40	52
Renegotiated a loan agreement or land contract to obtain lower interest rates*	19.8	4	20	45	44
Renegotiated a loan agreement or land contract to obtain a longer repayment period*	16.9	3	17	37	45
Changed lending institutions*	11.9	4	13	24	22

*Differences are statistically significant at the .05 level.

Source: Albrecht et al. 1987.

or 1985, and more than half had begun farming during the 1970s (Leistritz et al. 1987a). The characteristics of their farms were quite similar to those found in a statewide survey of farmers who were currently operating and considered farming to be their primary occupation. For example, the average number of acres operated by the displaced farmers was only about seven percent less than that operated by current farmers (Table 5.3). Both groups were dominated by commercial-size family farms with total sales in the range of $40,000 to $250,000, but the former farmers differed from their counterparts with respect to their age and level of education. Almost two-thirds of the displaced operators were less than age 45 (compared to less than half of the current operators), and more than half had attended college or another form of postsecondary school.

The former farmers were asked how they disposed of their assets. Only 14 percent of 142 former farmers sold all their land, 44 percent deeded their land back to a private individual or a financial institution, and another 27 percent retained ownership to all of their land (Figure 5.3). Several respondents reported that a combination of means were used. Over 75 percent of the former farmers sold their machinery either publicly or privately; just over 5 percent conveyed it back to a creditor, and about 6 percent retained it.

Overall, the former farmers had an average total debt of about $263,000 at the time they liquidated (Table 5.4). About 38 percent of these individuals were able to satisfy all of their obligations to their creditors. The remaining producers were not able to fully meet all their obligations. On average, these persons left about $123,300 in unpaid claims when they ceased farming (the median amount was $65,000). Of the total debt owed by the former farmers surveyed, about 30 percent was reported to have been unpaid. Nevertheless, only 11 percent of the respondents had filed for protection under bankruptcy laws.

Creditors have been affected substantially by the decapitalization of agriculture that is occurring today. The operators surveyed reported a total of 425 loans, of which 31 percent were operating loans, 24 percent were intermediate-term loans, and the remainder were long-term loans secured by real estate. Overall, 39 percent of all loans and unsecured debts were not paid in full when the farming operation was liquidated. The percentages varied substantially among lenders, ranging from 46 percent for FmHA to 35 percent for the Federal Land Bank, 29 percent for commercial banks, and 24 percent for the PCA. Unsecured creditors had a much worse experience, however; three-fourths of these obligations (such as accounts with input suppliers and rent due to landlords) were not paid in full.

Table 5.3

**Characteristics of Displaced Farmers and Current Farmers,
North Dakota, 1986**

Characteristics	Unit	Former Farmers	Current Farmers
Year started farming:			
Before 1950	Percent	13.0	17.0
1950–1954	Percent	3.0	11.1
1955–1959	Percent	7.1	11.4
1960–1964	Percent	8.9	11.1
1965–1969	Percent	11.2	9.8
1970–1974	Percent	22.5	14.8
1975–1979	Percent	29.0	15.7
1980–1984	Percent	5.3	8.9
Total acres operated:			
Mean	Number	1,466.0	1,556.9
Median	Number	1,220.0	1,200.0
Gross cash farm income (during last complete year of farming)			
Average	Dollars	101,045	110,266
Distribution:			
Less than $10,000	Percent	2.5	1.8
$10,000–$19,999	Percent	2.5	3.1
$20,000–$39,999	Percent	14.9	16.1
$40,000–$99,999	Percent	44.1	41.2
$100,000–$249,999	percent	26.1	30.4
$250,000 and over	Percent	9.9	7.4
Respondent's age:			
Less than 35 years	Percent	29.3	23.5
35 to 44 years	Percent	35.3	23.4
45 to 54 years	Percent	22.2	25.6
55 to 64 years	Percent	13.2	27.6
Respondent's education:			
Some high school or less	Percent	10.8	25.2
Completed high school	Percent	38.3	36.1
Attended college or other postsecondary school	Percent	36.5	26.5
Completed college	Percent	14.4	12.2

Source: Leistritz et al. 1987a.

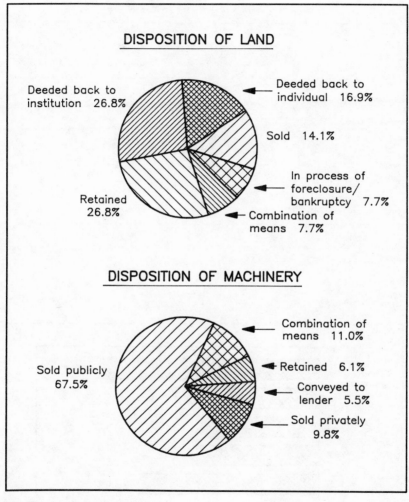

DISPOSITION OF LAND

Deeded back to
institution 26.8%

Deeded back to
individual 16.9%

Sold 14.1%

In process of
foreclosure/
bankruptcy 7.7%

Combination of
means 7.7%

Retained
26.8%

DISPOSITION OF MACHINERY

Combination of
means 11.0%

Retained 6.1%

Conveyed to
lender 5.5%

Sold privately
9.8%

Sold publicly
67.5%

SOURCE: Leistritz et al. 1987a

Figure 5.3 Methods of asset liquidation for land and machinery
for former North Dakota farmers

Table 5.4

Summary of Unpaid Liabilities by Type of Lender

Lender	Percent of Obligations Not Fully Satisfied to Source	Percent of Obligation Dollars Not Paid to Source	Percent of Total Loan Volume (Dollars)	Percent of Total Loan Dollars Not Paid
Banks	28.6	16.9	15.4	8.8
PCA	23.5	14.0	4.2	2.0
FmHA	45.7	37.4	40.9	51.5
Private individuals	20.0	29.9	0.4	0.4
Machinery companies	30.0	9.3	0.7	0.2
Federal Land Bank	34.7	32.4	19.1	20.8
Contract for deed	12.5	23.2	5.3	4.1
Other	2.4	4.3	9.7	1.4
Unsecured creditors	76.7	73.9	4.3	10.8
Total	38.8	29.7	100.0	100.0

Source: Leistritz et al. 1987a.

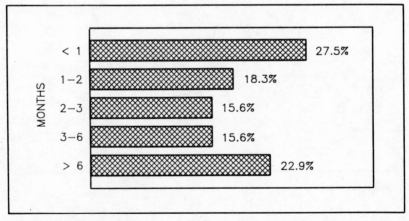

SOURCE: Leistritz et al. 1987a

Figure 5.4 Months required to find employment after leaving farming.

Unpaid liabilities represented substantial losses for some creditors. Overall, about 28 percent of the total value of the displaced farmers' total operating, intermediate-term, and long-term loans were not repaid when the farm operation was liquidated. Among secured creditors percentage losses were highest specifically on long-term loans and, overall, for the FmHA and Federal Land Bank. As stated earlier, unsecured creditors also suffered high percentage losses.

At the time of the survey, about 55 percent of the respondents were still living in the county where their farm had been located, 32 percent had relocated to another county in the state, and 13 percent had moved out of state. Respondents under age 45 were more likely to have moved from their home county. Similarly, those with completed college degrees were much more likely to have moved from their home county, although there were no marked differences in relocation among those with other educational levels. Relocation status did not seem to be consistently related to the year that the operator ceased farming; however, there was a slight tendency for respondents with higher levels of net worth to remain in the community, while those with lower levels of equity chose to relocate.

Finding suitable employment has been identified as one of the most frequently encountered problems associated with the transition out of farming (Graham 1986; Hill 1962; Guither 1963). The survey respondents were asked how many months it had taken to find employment. About 46 percent of the respondents reported a search of 2 months or less, and slightly over 60 percent searched for less than 3 months (Figure 5.4). About 41 percent indicated that they had to move to find employment, but about 58 percent of these would have preferred to have stayed in or near their hometown.

The experience of these displaced farm operators in seeking alternative employment can be compared to that of displaced workers nationwide. A 1984 survey of more than 5 million workers who had lost their jobs after a tenure of three or more years revealed that, of the 60 percent who had found new employment at the time of the survey, the median period without work had been 6 months, and 14 percent had relocated to find work (Flaim and Sehgal 1985). In contrast, about 83 percent of the displaced farmers had found employment, the median period without work had been slightly over 2 months, but 41 percent of these had to move to find a job.

In addition to the 83 percent of the respondents employed at the time of the survey, 8 percent were unemployed, 6 percent were full-time students, and about 3 percent were retired. The percentage of former farmers who were currently employed was highest (95 percent) for those who were under age 45 and had no education beyond high school. For those over age 45 with no education beyond high school, however, the unemployment rate was nearly 23 percent. Overall, 57 percent of their spouses were currently employed, but the percentages varied considerably by educational level. Of those who were under age 45 and had completed college, 83 percent were employed. On the other hand, one half of the spouses with no education beyond high school were not employed. The occupations most frequently reported by respondents were farm work (17.8 percent), sales (15.5 percent), transportation (14.0 percent), and construction crafts (12.4 percent), while their spouses were most often employed in administrative support (36.0 percent), professional specialities (19.8 percent), services (16.3 percent), or sales (10.5 percent).

About 71 percent of the employed respondents indicated that they were satisfied with their present employment. However, 39 percent of all respondents indicated that they were likely to look for different employment in 1986, and about 31 percent of all spouses were described as likely to look for different employment. The respondents would most often seek employment as truck drivers or farm workers, and spouses were most likely to seek jobs as bookkeepers, registered nurses, and secretaries.

Among the households where one or both marriage partners were likely to look for a different job, about 70 percent indicated that they were willing to move to another community to find employment. The fact that about 27 percent of all survey respondents fell into this category suggests that the transition process may not yet be complete for many displaced farm families.

Studies conducted in Iowa (Otto 1985) and Ohio (Henderson and Frank 1986) led to conclusions similar to those of the North Dakota analysis in most respects. Both found that most of the displaced farmers had been operating commercial-size family farms and that they were generally younger than the average current producer. Assets were typically liquidated through voluntary sale or conveyance to creditors rather than through foreclosure or bankruptcy. For example, only 13.7 percent of Ohio farmers studied had filed for bankruptcy. Most of the displaced producers had succeeded in finding alternative employment; only 9.3 percent of the Ohio farmers and 14.9 percent of those in Iowa were unemployed at the time the studies were conducted.

One aspect in which the results of the Iowa and Ohio studies differed from those of the North Dakota survey was the relocation of displaced farmers. The percentage of displaced North Dakota farmers who had relocated from their home counties (45 percent) was substantially higher than corresponding values in Ohio (27 percent) and Iowa (23 percent). One likely explanation for the difference would be the sparse population and small number of nonfarm job opportunities in many of North Dakota's more agriculturally dependent counties. In North Dakota's 39 agriculturally dependent counties (as defined by Bender et al. 1985), total employment declined 6.2 percent from 1980 to 1985 (Leistritz et al. 1987a).

Conclusions and Implications

The purpose of this chapter was to examine adaptations that farm households are making in response to current economic conditions. These adaptations include changes in farm management practices, farm and financial organization, family lifestyles, and farm liquidation. Surveys of farmers in a number of states indicate that substantial percentages are making changes in their management practices and lifestyles. Postponing capital expenditures (e.g., for new machinery) and reducing family living expenses were adaptations reported by many farmers. Substantial numbers also had attempted to reduce their farm operating costs in areas such as tillage, fertilizer, and chemicals. Debt and asset restructuring were also being practiced by significant numbers of heavily indebted farmers. For example, 18 percent of North Dakota farmers had renegotiated a loan during 1985 to reduce interest payments.

For some farmers, however, changes in management practices and/ or reorganization of debt and assets have not been adequate strategies. For these producers, liquidation of the farm business has become the only viable alternative. Studies of farm liquidations conducted in several states have indicated that these operators tend to be younger than average and to have commercial-size family farms. Assets have typically been liquidated through voluntary sale or conveyance to creditors rather than through foreclosure or bankruptcy. Most of the displaced farmers had succeeded in obtaining alternative employment, but substantial numbers had relocated from their home county to find work.

The adaptations of producers to the conditions of economic stress in agriculture may have major implications for rural, farm-dependent communities. Farm households' reductions in living expenses and postponement of capital purchases could result in substantial reductions in sales for local merchants while the relocation of displaced farm families may substantially affect the population profile of rural areas. These secondary effects of farm financial stress are examined in the next chapter.

6

Demographic Characteristics of Rural Residents in Financial Distress and Social and Community Impacts of the Farm Crisis

Steve H. Murdock, Rita R. Hamm,
Lloyd B. Potter and Don E. Albrecht

The preceding chapter examined producers' responses to the farm crisis. In this chapter we examine the demographic characteristics of those being impacted by the crisis and the social and community-level impacts of the crisis. The intent is to address such questions as: What are the characteristics of those who are most likely to fail in farming and in other business enterprises in rural areas as a result of the current problems in agriculture? What are the age, educational experience, and other characteristics of such persons? How do their characteristics differ from those who are remaining on the farm or in rural business enterprises? What are the perceived and reported impacts of the crisis on the social and personal lives of rural residents? What are the impacts of the crisis on rural community businesses and services? The answers to such questions are instrumental for discerning who has been impacted and to what extent, and they serve as an essential base for projecting the long-term implications of the crisis for rural America.

Research related to the characteristics of persons being impacted and about the social and economic impacts of the crisis for producers is accumulating (for example, Heffernan and Heffernan 1985a, 1985b; Bultena et al. 1986; Murdock et al. 1986a). Such analyses have tended to show that those producers most likely to have high debt-to-asset ratios are young, well educated, and innovative producers who are managing commercial-scale farms. Several analyses have noted producers' perceptions of the impacts of the crisis on their farms and on their personal lives (Heffernan and Heffernan 1985a; Albrecht et al. 1987b;

Leistritz et al. 1987b). For example, the Heffernan's found that more than 97 percent of their respondents had suffered depression due to the crisis, and more than 60 percent withdrew from family and friends (Heffernan and Heffernan 1985a). Albrecht et al. (1987b) also found that a substantial, although lower, proportion of producers had experienced depression and marital problems, and both Albrecht et al. (1987b) and Bultena et al. (1986) found that the extent of stress was related to the degree of indebtedness. Significantly, both Albrecht's and Bultena's efforts were based on random sample surveys of producers in their respective states (Texas and Iowa).

Despite some progress in addressing such issues, however, numerous gaps exist in the literature. Foremost among these is the fact that very few studies have been longitudinal in nature (however, see Albrecht et al. 1987a, 1987b; Leistritz et al. 1987c). In addition, there is very little information on the crisis' impacts on rural residents who are not farmers and ranchers, and the few existing analyses tend to be case studies of single communities (Murdock et al. 1987b; Doeksen 1987). Data comparing the perceptions of producers and other rural residents and providing comparisons among persons who have been impacted to different degrees by the farm crisis have seldom been available.

In the analysis reported in this chapter, we address the questions noted above by examining the characteristics of two sets of persons being impacted by the crisis—producers and residents in agriculturally dependent rural communities, and among each we examine groups that have been directly or indirectly impacted by the crisis. Specifically, we examine the responses of producers with different levels of debt, ones who have been forced to leave farming due to the crisis, and former business persons who have been forced to leave business. For each group, we first examine their demographic characteristics. We then examine their perspectives on the crisis, on its impacts on their personal lives, and on the businesses and services in their communities.By so doing, it is possible to more firmly establish the characteristics of those who are being most directly impacted by the crisis and to discern the characteristics of such impacts.

In describing the characteristics of those being affected by financial stress, we examine data from several different data sets. These include our 1985 and 1986 surveys of farmers in Texas and North Dakota, the 1986 survey of former farmers in North Dakota, and the community surveys of businesses and residents in North Dakota and Texas (details on the collection of these data sets are presented in the Technical Appendix). Because of the importance of levels of financial difficulty in influencing producers' responses, data for producers are shown for those in different debt-to-asset categories. After examining the results of the analysis from these data sets, we present generalizations concerning the characteristics and the impacts of the farm crisis on producers and other rural residents.

Demographic and Socioeconomic Characteristics of Producers, Business Operators, and Other Rural Residents Under Financial Stress

Among the most pervasive characteristics of producers under stress is their relatively young age. Nationally, more than 20 percent of those less than 35 years of age had debt-to-asset ratios of more than 70 percent in 1986 compared to 12 percent of those 35 through 44, 9 percent of those 45 through 54, 3 percent of those 55 through 64, and 1 percent of those 65 years of age or older (Johnson et al. 1986). Table 6.1 shows similar age characteristics among the combined sample of respondents from the North Dakota and Texas surveys of farm operators. A review of the data in this table clearly indicates that the producers with the highest debt-to-asset ratios are those who are in the young adult age groups. In this combined sample of North Dakota and Texas producers, more than 56 percent of those with debt-to-asset ratios of 41 to 70 percent were 25 through 44 years of age, and of those with debt-to-asset ratios of more than 70 percent, nearly 50 percent were in these age groups. On the other hand, only 15 percent of those with debt-to-asset ratios of 41–70 percent and 21 percent of those with debt-to-asset ratios exceeding 70 percent were in the age group of 55 through 64 years of age. For those with no debt, only 15 percent were less than 45 years of age, and nearly 61 percent were over 55; for those with debt-to-asset ratios of 1–40 percent, 39 percent were under 45 and 32 percent were more than 55 years of age.

In like manner, those experiencing higher levels of financial stress (as indicated by debt-to-asset ratios exceeding 40 percent) were likely to have begun farming during the 1970s. Of those who had debt-to-asset ratios of 41–70 percent, 41 percent began farming during the 1970s, while among those with ratios of greater than 70 percent, 42 percent began farming between 1970 and 1979. Of those with debt-to-asset ratios of 1–40 percent, only 27 percent began farming during the 1970s (59 percent began farming between 1945 and 1964), while for those with no debt, only 14.6 percent had begun farming during this time period (55 percent began in the 1945–1964 time period).

Those with the highest debt-to-asset ratios are also relatively well educated; more than 48 percent of those with debt-to-asset ratios of 41 to 70 percent had at least some college, and over 41 percent of those

Table 6.1

**Selected Characteristics of Farmers and Farm Operations
by Debt-to-Asset Ratio Categories**

Characteristics	Debt-to-Asset Ratio[1]				Total Sample
	No Debt	.01-.40	.41-.70	> .70	
Age of respondent					
< 25	1.2	2.6	5.0	6.8	3.4
25-34	6.4	11.2	26.0	26.2	15.1
35-44	8.0	25.3	30.7	23.6	22.8
45-54	23.9	28.4	23.5	22.8	25.9
55-64	60.6	32.4	14.9	20.7	32.9
Mean	53.8	47.4	41.4	42.1	47.3
S.D.[2]	9.8	10.8	10.6	11.7	11.4
Year started farming					
1980-1984	6.1	7.7	8.4	11.3	8.0
1975-1979	7.0	11.8	22.0	22.7	14.3
1970-1974	7.6	15.4	19.2	19.3	15.2
1965-1969	7.0	10.2	12.7	9.7	10.0
1955-1964	18.0	25.2	23.2	17.2	22.4
1945-1954	36.9	24.1	12.4	16.4	23.3
1920-1944	17.4	5.7	2.2	3.4	6.9
Respondents' formal education					
Did not complete high school	30.3	20.1	18.4	20.2	21.7
Completed high school	37.6	38.1	32.9	38.6	37.1
Some postsecondary education	21.7	24.1	30.1	28.3	25.4
Four (or +) years of college	10.5	17.7	18.7	12.9	15.9
Household/family					
Mean no. persons/household	2.7	3.2	3.7	3.5	3.2
Mean no. children < age 18	0.4	1.0	1.3	1.3	0.9
Net worth statements used	27.4	50.1	67.0	68.1	51.5
Profit/loss statements used	38.4	54.0	65.0	65.1	54.6
Cash flow statements used	29.8	43.6	58.5	58.2	45.8
Attendance at Extension Service meetings/programs	53.1	63.8	65.9	65.1	62.3
Extension Service/Experiment Station publications used	40.2	51.6	54.5	53.4	50.2
Off-farm employment					
Operator					
Employed off-farm (1984)	20.4	24.8	27.9	29.4	25.2
Plans off-farm work (1985)	2.2	7.0	13.8	17.1	8.6
Years of full-time, nonfarm employment					
Mean	8.7	6.6	4.6	5.3	6.9
S.D.	12.6	9.6	6.7	7.6	10.0

Table 6.1 (Continued)

Characteristics	Debt-to-Asset Ratio[1]				Total Sample
	No Debt	.01–.40	.41–.70	> .70	
Spouse					
Employed off-farm (1984)	26.3	37.0	36.4	36.3	34.8
Plans off-farm work (1985)	3.3	4.3	12.4	14.0	6.9
Off-farm income:					
Mean ($)	5,256	8,873	7,180	10,064	7,898
S.D. ($)	11,953	17,621	15,399	21,460	16,859
Community organizations					
Farm					
Member	67.0	71.6	67.8	64.7	69.0
Officer	8.3	12.3	10.8	10.0	10.9
Commodity					
Member	24.2	26.9	30.7	23.4	26.6
Officer	6.3	8.6	8.6	5.3	7.1
Civic club					
Member	24.5	29.8	36.2	33.6	30.2
Officer	16.2	13.7	11.8	14.6	13.9
PTA					
Member	15.5	24.7	25.7	24.7	22.9
Officer	9.1	11.6	8.7	8.1	10.0
County commission					
Member	17.0	21.5	22.9	18.3	20.4
Officer	29.1	30.7	24.1	19.3	27.1
Church					
Member	89.7	90.1	92.3	94.0	90.9
Officer	27.5	29.3	27.8	29.3	28.7

[1]Percentages may not sum to 100 percent due to rounding.
[2]SD = standard deviation.

with ratios of more than 70 percent had such educational levels. For those with no debt, only 31 percent had attended college.

Similarly, those with high levels of debt have tended to be users of such financial management techniques as net worth statements, profit/loss statements and cash flow statements and to use such information sources as the Agricultural Extension Service and Experiment Stations. More than 68 percent with debt-to-asset ratios of more than 70 percent used net worth statements, more than 65 percent in this category used profit/loss statements, more than 58 percent used cash flow statements, 65 percent attended extension meetings, and 53 percent used publications from the extension and experiment stations compared to 27, 38, 30, 53, and 40 percent, respectively, for those with no debt. When combined with the findings on education, these data suggest that those producers with the most severe debt-to-asset ratios are not less innovative nor less well educated than those with lower levels of debt, and in fact, they tend to have higher levels of education and to be more likely to use financial management techniques than those with lower debt levels.

The fact that those producers with the highest debt-to-asset ratios tend to be those who are young adults in family-rearing ages is evident in the data on household and family size in Table 6.1. The mean size of households for those in the 41–70 percent debt-to-asset category was 3.7 and in the more than 70 percent category 3.5, compared to 2.7 in the no debt category.

An analysis of the data in Table 6.1 also suggests that many producers have used off-farm employment as one means of obtaining additional income. As an examination of the last data item in this section of the table indicates, the level of off-farm employment increases by nearly 10 percent from the no debt category to the highest debt-to-asset ratio category, and plans of the operator to seek work increase from 2 to 17 percent.

Finally, the last item in Table 6.1 shows the involvement of producers in community organizations, both as members and as officers of such organizations. An examination of the data in this table shows that producers in all debt-to-asset categories are active participants in community organizations. Thus, more than 67 percent of producers in all categories are members of farm organizations, more than 25 percent are members of commodity organizations, nearly one-third are members of civic clubs, nearly 25 percent are members of PTA, and nearly 90 percent are members of churches. In addition, in all categories of organizations producers are likely to be officers. Overall, these data on participation in community organizations suggest that producers do actively participate in their community and that those in the high debt-to-asset categories are as likely as those in other categories to be active participants and officers in such organizations. Clearly a loss of a substantial proportion of these producers would effect membership and leadership in community organizations.

Overall, the data for the states of Texas and North Dakota clearly verify those findings from other analyses (Bultena et al. 1986; Jolly et al. 1986; Heffernan and Heffernan 1985a), which show those producers having the highest debt-to-asset ratios to be young with established families who entered farming during the 1970s and who are relatively innovative. It is evident that such producers are not the marginal producers who have often failed in the past. It is also apparent that these producers are active members of their communities, often both participating in and leading community organizations.

Yet another means of examining the characteristics of those likely to fail in farming is to look at those who have already failed. In the present analysis, we examine the characteristics of the former farmer survey previously described. Although there are admittedly problems with the representativeness of this data set (see the Technical Appendix), it represents the largest data set known to these authors on the characteristics of those who have already failed in farming. Its uniquely appropriate focus for examining the questions of interest here justify its use in the present analysis.

Table 6.2 provides data on the characteristics of persons who are former farmers. An examination of the data in this table suggests that those failing in farming are similar to those who the data in Table 6.1 indicated had the highest debt-to-asset ratios. That is, of the former farmers more than 62 percent were 25 through 44 years of age at the time they quit farming, 52 percent began farming in the 1970s, 50 percent had some postsecondary training, and their average household size was more than 4 persons. Finally, it is evident that during their final year of farming, those producers who left farming made extensive use of off-farm employment in an attempt to meet expenses. During the final year of the operation of their farms, nearly 47 percent of producers and nearly 60 percent of the spouses were involved in off-farm employment (compared to 25 percent of producers currently farming and 35 percent of their spouses; see Table 6.1). An evaluation of these data suggests that those actually failing in farming are similar to those with high levels of debts-to-assets and are persons who are young and well-educated and who began farming during the 1970s.

The final set of data examined in this section comes from the community surveys in North Dakota and Texas. In these surveys current business operators, operators who had recently ceased operating businesses (for reasons other than retirement), and residents who were not involved in farming and who were not owners or managers of businesses were

Table 6.2

**Selected Characteristics of North Dakota
Farmers Who Had Left Farming
Due to the Farm Crisis**

Characteristic	Percent
Age of respondent[1]	
< 25	3.6
25-34	27.5
35-44	35.3
45-54	21.6
55-64	12.0
Mean	40.8
Standard deviation	10.9
Year started farming	
1980-1984	5.4
1975-1979	29.4
1970-1974	22.8
1965-1969	11.4
1955-1964	16.2
1945-1954	10.2
1920-1944	3.6
Mean	19.7
Standard deviation	12.5
Respondents' formal education	
Did not complete high school	16.8
Completed high school	38.3
Some postsecondary education	36.5
Four (or +) years of college	14.4
Household/family	
Mean no. persons/household	4.1
Mean no. persons < age 18	1.8
Employed off-farm one year	
prior to leaving farming	
Operators	46.7
Mean no. of years of	
full-time off-farm work	4.9
Spouses	59.6

(N = 167)

[1]Age when operator ceased farming.

interviewed. The data in Table 6.3 show information on the age, education, and family characteristics of such persons in nine communities.

An evaluation of the data in Table 6.3 suggests that the characteristics of business operators being forced to cease operation are not significantly different than those of other business operators who have remained in business in regard to age or household size, but they do differ in regard to the number of years they have been in business. The mean number of years in business was 8.6 for former business operators and 12.9 for current business operators. In addition, 56 percent of former business operators had been in business for less than 5 years at the time they ceased operating their businesses, compared to 30 percent of current business operators. Former and current business operators have characteristics that are different from those of other community residents, suggesting that business operators possess educational and other characteristics that should allow them to adapt more successfully to changing economic conditions. The fact that those whose businesses have recently failed are of the same ages and educational levels and have other characteristics similar to those who have remained in business also suggests that those who have failed in business are not marginal operators but different only in the fact that a much higher proportion started business during the most recent period in which business activity has been insufficient to provide profitability. Similar to agricultural producers, then, it appears that failure in rural businesses has been closely linked to a unique period of time with specific characteristics (such as high interest rates and high capital investment requirements).

That current business operators are, in fact, taking steps to alleviate the impacts of the crisis on their businesses was evident from the results of the North Dakota and Texas surveys of this group. The survey data (not shown) indicated that a substantial proportion of current business operators have taken steps such as restructuring debt (19.7 percent), reducing inventories (50.6 percent), increasing collection efforts (67.1 percent), and reducing labor forces (41.7 percent) in order to maintain the profitability of their businesses. As with producers, business operators have been forced to alter their business practices substantially in order to survive the crisis.

Overall, an examination of the data in this section suggests that the farm crisis is impacting producers and business operators who are in their most productive ages. They are not persons who are at retirement ages and thus ones that have voluntarily reduced production, or persons who have low educational levels that have put them at a disadvantage in competing for sales or business, but persons who, under normal economic conditions, might have been expected to be successful. In sum, the data in this section suggest that the major characteristic related

Table 6.3

**Selected Socioeconomic Characteristics of Current Business Operators,
Former Business Operators and Community Residents
in North Dakota and Texas Communities, 1986**

Characteristics	Current Business Operators	Former Business Operators	Community Residents
Age of respondents			
< 25	4.5	4.8	8.8
25-35	22.6	20.5	25.7
35-44	32.1	30.1	24.0
45-54	22.4	24.1	20.7
55-64	18.5	20.5	20.7
Mean age of respondents	43	43	42[a]
$(X^2 = 24.1^c)$			
Respondents' formal education			
Did not complete high school	6.8	9.8	13.3
Completed high school	27.7	26.8	37.9
Some post-secondary education	31.6	40.2	28.9
Completed 4 (or +) years college	33.9	23.2	19.9
$(X^2 = 62.3^c)$			
Mean age of spouse	42	42	42*
Average mean household size	3.3	3.1	3.0[a]
Mean number of children under 19	1.3	1.2	1.5[ab]
Mean number of years of residence in city	20.8	18.5	19.3
Years in business			
1-5	30.2	55.7	---
6-10	23.8	17.7	---
11-20	24.7	17.7	---
21-30	12.7	3.8	---
31-62	6.9	5.1	---
Mean	12.9	8.6*	---
S.D.	10.8	9.2	---
N =	714	83	829

Note: [d]Would indicate that values for current and former business operators were significant, but no significant differences were found.
[a]Means for current business operators and community residents are significantly different at the .05 level.
[b]Means for former business operators and community residents are significantly different at the .05 level.
[c]Chi square statistic is significant at .05 level.
*Means are not significantly different at the .05 level.

to failure has been the misfortune of starting in business during a period with high interest costs and property values that was followed by a period of rapid deflation and decline in prices, values, and business volume.

Producers' and Rural Residents' Perceptions of the Farm Crisis

Before examining the impacts of the crisis on the social and community characteristics of rural areas, it is useful to discern how the crisis is being perceived by rural residents. Is the crisis seen as a function of general economic and market conditions beyond the control of producers or as being due to failures in the actions of producers? What are seen to be the major causes of the crisis? How favorably are farmers and farming generally perceived by different groups in rural areas?

Table 6.4 presents data on responses by producers to questions related to farming in general, regarding the specific causes of the farm crisis, the need and desirability of financial assistance for farmers, and concerning producers' perceptions of the financial stability of their own farms and their satisfaction with farming. An examination of the data in the first panel of Table 6.4 indicates widespread agreement among producers that agriculture is the most basic industry in the economy, that farmers should organize to bargain more effectively, that a larger proportion of farmers are in financial trouble than in previous time periods, that farmers will always be able to supply enough food to feed America, and that large corporations control agriculture. They show similar and high levels of disagreement with the premise that farmers are to blame for high food prices.

Substantial differences do exist among those in different debt-to-asset categories for several other factors, however. That is, those with high debt-to-asset ratios tend to be less likely than those with lower debt levels to believe that farms are too large, that farming is strictly a business, and that new forms of technology will assist farmers in surviving. They are more likely to believe that the family farm is going out of existence. Overall, an evaluation of the data in this panel of Table 6.4, indicates that producers see agriculture as the backbone of the American economy and as being impacted by corporate actions. Their financial position appears to affect their perspectives on the business nature of farming, on farm size, on the utility of technology, and on the continued existence of the family farm. It thus appears that, although the crisis has not altered producers' perceptions of the central role of agriculture in the U.S. economy, those most severely impacted are more

Table 6.4

Responses of Texas and North Dakota Producers to Selected Questions Concerning the General Conditions in Farming, the Causes of the Farm Crisis, Assistance for Financially Troubled Farmers and Perceptions of the Status of Own Farm and Satisfaction With Farming

Selected Questions	Chi Square	Debt-to-Asset Ratio				Total Sample
		No Debt	.01-.40	.41-.70	> .70	
Perceptions of Conditions/Policies		(percent* agreeing/strongly agreeing)				
Most farms today are too large	20.5	29.7	26.9	23.2	20.0	25.8
Farmers should organize to bargain for the prices of farm products	10.5[a]	76.0	78.6	76.4	78.3	77.7
The proportion of farmers who are now in financial trouble is much greater than at most times in the past	3.3	90.0	94.3	96.3	92.9	93.7
Agriculture is our nation's most basic industry	10.9	98.3	97.4	98.1	94.7	97.4
To me, farming is strictly a business	21.1	82.6	73.8	69.2	71.4	74.4
The family farm is rapidly going out of existence	33.2	69.0	74.7	81.2	79.0	75.2
Agriculture plays a vital role in the nation's economy	9.4[a]	98.6	98.4	98.5	97.7	98.4
American farmers will always be able to produce enough food to feed America	14.6[a]	71.3	76.3	78.7	75.3	75.6
Today, large corporations, not farmers, control agriculture	13.8[a]	56.4	57.8	60.2	59.1	58.2
Farmers are primarily to blame for high food prices	9.8[a]	2.0	1.7	1.9	1.2	1.7
Farmers who adopt new technology and production methods will survive	13.3[a]	68.4	62.0	57.3	52.8	62.2
Agreement with Specific Policy Alternatives						
The only solution to the farm problem is mandatory production controls (or marketing quotas)	24.7	39.0	46.4	54.0	65.4	47.1
The government should phase out making direct payments to farms	55.2	60.8	51.4	33.7	27.8	49.5
We should not export our farm products to foreign countries if they are not willing to pay what it costs us to produce them	22.0	76.5	69.9	69.1	58.5	70.7
U.S. agriculture has become too dependent on international trade	15.8[a]	58.8	62.5	67.8	64.2	62.4
Perceived Reasons for the Crisis		(- - - - -percent indicating cause was important/somewhat important- - - - -)				
Farmers' paying too much for land or equipment**	12.8	95.1	94.6	90.1	91.1	93.9
Farmers' not participating in farm programs**	4.6[a]	54.2	60.5	62.7	50.0	58.2
Low farm yields**	2.1[a]	65.2	68.4	69.3	66.0	67.5
Farmers' making poor marketing decisions**	8.1[a]	70.0	73.9	68.2	58.9	70.8
Farmers' not using up-to-date technology**	15.1	57.2	61.6	53.9	57.1	59.1
Bad weather**	7.5[a]	84.0	88.9	85.6	89.3	87.2
High interest rates	22.1	92.4	95.7	98.5	96.5	95.6
Low prices for farm products	4.8[a]	98.0	98.7	98.9	99.5	98.7
Government involvement in agriculture	8.0[a]	87.0	89.8	89.8	85.9	88.8
Corporate farms	17.6	45.0	46.4	35.7	36.5	42.9
Farmers' attempting to expand the size of their farms too rapidly	53.0	87.7	91.0	79.0	75.0	86.2
Farmers' being poor managers	67.3	87.7	84.2	67.4	68.1	79.9
The high cost of farm supplies and equipment	4.8[a]	96.7	98.1	97.0	97.1	97.5
Changing land values	6.5[a]	85.2	87.8	86.8	87.6	87.0
Changing export markets for farm products	29.9	88.5	95.6	94.3	94.1	93.7
Farmers' living beyond their means	36.8	86.7	83.4	73.8	70.9	80.7
Reasons Given as Most Important		(- - - - - percent indicating reason is most important - - - - -)				
Low farm prices		40.9	43.3	53.9	46.4	44.4
Government involvement in agriculture		10.8	9.6	13.5	19.6	11.3
High interest rates		3.4	8.7	10.1	7.1	7.4
High cost of farm supplies and equipment		11.4	4.9	0.0	0.0	6.5
Farmers pay too much for land and equipment		6.8	5.2	2.2	1.8	5.0
Changing export markets		4.0	6.7	0.0	1.8	5.4

Table 6.4 (Continued)

Selected Questions	Chi Square	No Debt	.01-.40	.41-.70	> .70	Total Sample
		Debt-to-Asset Ratio				

Selected Questions	Chi Square	No Debt	.01-.40	.41-.70	> .70	Total Sample
Financial Assistance for Producers		(– – – – – – percent – – – – – –)				
Agreement With Need for Federal Aid						
(1) Strongly agree		89.3	2.7	5.0	7.4	15.5
(2) Agree		24.2	31.9	43.4	39.9	33.3
(3) Neither agree nor disagree		20.8	22.5	19.8	19.0	21.3
(4) Disagree		35.5	31.6	25.6	20.2	29.9
(5) Strongly disagree		16.7	9.0	3.9	5.4	9.2
Type of Federal Aid		(percent agreeing/strongly agreeing)				
Providing financial assistance to financially troubled ag creditors either directly or through loan guarantees	2.5[a]	84.1	87.0	89.5	82.6	86.5
Subsidizing interest rates on operating loans	13.1	74.6	71.3	85.7	84.9	77.5
Participating with creditors and farmers in restructuring land debt	4.9[a]	82.4	84.3	90.8	89.8	86.6
Providing low-interest loans or grants to financially stressed farm families for vocational training or college in preparation for a new occupation	3.1[a]	85.3	89.1	84.6	82.4	86.4
Agreement With Need for State Aid		(– – – – – – percent – – – – – –)				
(1) Strongly agree	82.0	1.4	3.0	3.5	10.7	3.7
(2) Agree		17.1	21.3	33.7	32.7	24.1
(3) Neither agree nor disagree		15.8	22.1	19.4	14.9	19.4
(4) Disagree		46.2	41.9	34.5	34.5	40.5
(5) Strongly disagree		19.5	11.8	8.9	7.1	12.3
Type of State Aid		(percent agreeing/strongly agreeing)				
Providing financial assistance to financially troubled ag creditors either directly or through loan guarantees	1.6[a]	81.2	83.2	88.0	82.1	84.0
Subsidizing interest rates on operating loans	10.5	72.9	70.3	84.4	85.9	77.1
Participating with creditors and farmers in restructuring land debt	9.4	82.6	81.6	93.5	91.5	86.7
Providing low-interest loans or grants to financially stressed farm families for vocational training or college in preparation for a new occupation	1.9[a]	92.3	86.8	89.2	91.7	89.1
Perception of Financial Status/ Satisfaction With Farming						
Considering your farm's overall financial situation, how likely is it that you will expand your operations in the next three years?	35.6	(– – – – – – –percent– – – – – –)				
(1) Very likely		6.7	7.7	5.6	3.6	6.6
(2) Likely		8.8	16.5	15.4	11.2	14.1
(3) Don't know		8.8	12.6	17.3	8.9	12.2
(4) Unlikely		25.9	25.4	20.7	27.2	24.8
(5) Very unlikely		49.8	37.7	41.0	49.1	42.2
Considering your farm's overall financial situation, how likely is it that you will be able to continue to farm for at least the next three years?	129.5					
(1) Very likely		50.3	37.5	18.7	17.0	34.2
(2) Likely		32.2	42.2	42.7	35.7	39.5
(3) Don't know		11.1	14.5	25.5	29.8	17.7
(4) Unlikely		2.7	3.2	7.5	9.9	4.7
(5) Very unlikely		3.7	2.6	5.6	7.6	4.0
Compared to other farmers in your area, do you feel your financial situation is:	29.5					
(1) Above average		48.0	45.1	20.2	29.1	41.3
(2) About the same		50.3	49.7	73.0	61.8	54.0
(3) Below average		1.7	5.2	6.7	9.1	4.8

Table 6.4 (Continued)

Selected Questions	Chi Square	Debt-to-Asset Ratio				Total Sample
		No Debt	.01–.40	.41–.70	> .70	
How satisfied are you with the current financial returns in farming? Are you:	72.1					
(1) Completely satisfied		0.7	1.3	0.0	0.6	0.8
(2) Satisfied		13.4	8.4	3.4	1.2	7.6
(3) Neither satisfied nor dissatisfied		5.7	6.8	5.2	1.8	5.7
(4) Dissatisfied		49.7	43.8	40.1	37.4	43.6
(5) Completely dissatisfied		30.5	39.6	51.3	59.1	42.2
How satisfied are you currently with farming as an occupation? Are you:	28.8					
(1) Completely satisfied		32.9	28.1	20.6	18.7	26.6
(2) Satisfied		45.3	50.3	51.3	48.5	49.3
(3) Neither satisfied nor dissatisfied		7.7	4.9	9.4	10.5	7.0
(4) Dissatisfied		10.1	11.4	13.5	14.6	11.9
(5) Completely dissatisfied		4.0	5.2	5.2	7.6	5.3
All things considered, how satisfied are you with farming at the present time? Are you:	68.6	(– – – – – – – –percent– – – – – – –)				
(1) Completely satisfied		11.2	4.0	2.2	0.0	5.3
(2) Satisfied		40.2	40.1	32.2	18.2	37.3
(3) Neither satisfied nor dissatisfied		9.5	8.9	10.0	7.3	9.1
(4) Dissatisfied		27.4	33.8	41.1	58.2	35.1
(5) Completely dissatisfied		11.7	13.2	14.4	16.4	13.2

Note: Chi squares were calculated from contingency tables of responses for each variable by the four respondent–debt categories. Percentages were calculated within respondent categories.
[a]Chi square statistic **not** significant at .05 level.
*Percents may not total 100, due to rounding.
**These questions were asked only of Texas producers.

likely to see agriculture as vulnerable to long-term difficulties that cannot be alleviated by such traditional factors as the use of new technology.

The second panel in Table 6.4 examines producers' perceptions of the reasons for the crisis. An examination of the data in this table indicates that producers in all debt-to-asset categories agree that land and equipment costs, high interest rates, low commodity prices, changes in land values, and changes in export markets have been major causes of the farm crisis. Those with high levels of debts relative to assets, however, are less likely to see actions of producers, such as poor marketing conditions, attempts to expand too rapidly, poor management, and farmers living beyond their means as major causes of the financial condition of producers. It is apparent that those experiencing negative impacts (i.e., those with higher debt-to-asset ratios) are more likely to blame circumstances rather than actions by producers, perhaps because of the self blame that is implied by agreement that actions of producers may have played a role in their financial problems.

The third panel of Table 6.4 examines producers' levels of support for federal and state aid to financially stressed farmers in general and for specific types of potential aid. The data in this table show that levels of debt relative to assets affect support for general federal and state aid. Whereas more than 50 percent of those with no debt disagree with the assertion that financially troubled farmers should receive federal aid and more than 65 percent oppose state aid, only 25 percent of those with debt-to-asset ratios of greater than 70 percent oppose federal aid and only 41 percent oppose state aid. Among those agreeing with the need for aid, however, those with different debt-to-asset ratios disagree very little in regard to the desirability of having the federal and state government provide financial assistance to agricultural creditors through loan guarantees, subsidizing interest rates on operating loans, restructuring land debt, or providing financial assistance to producers in preparing for new occupations. More than 70 percent of those in all debt-to-asset categories supported such forms of aid. Finally, producers in Texas were also asked to evaluate four key policy issues. The results of this analysis suggest that those with higher levels of debt were much more likely to support mandatory production controls and direct government payments to producers than those with little or no debt.

The final panel of Table 6.4 provides data on responses to producers' evaluations of the financial condition of their own farm and their satisfaction with farming. The results in this table indicate that those under the most severe financial stress are more likely to perceive that their chances of remaining in farming are poor and to show higher levels of dissatisfaction with the financial returns in farming and with farming overall. It is interesting to note, however, that the large majority

of producers in all debt-to-asset categories are satisfied with the occupation of farming. It appears that the current financial crisis has
affected the levels of satisfaction and the perceptions of the future of
those experiencing the greatest financial stress but has not caused them
to reject the occupation of farming as a desirable career.

The data in Table 6.4 suggest that producers perceive, overall, that
agriculture is the most important industry in the United States; that
major economic factors, such as export markets, poor prices, high interest
rates, and similar factors, are the major causes of the crisis; that financial
assistance should generally not be provided by either the federal or
state government to aid financially troubled farmers; and that farming
is generally a desirable occupation. On the other hand, the results show
blame displacement and disenchantment among those producers experiencing the most severe financial stress. Such producers see the major
causes of the crisis as being events beyond their control and show a
substantially higher level of support for federal and state assistance than
those with lower levels of debt relative to assets.

Table 6.5 presents data for former farmers regarding the causes of
the crisis and the degree to which federal and state aid is appropriate.
The top panel of the table clearly suggests that former producers have
especially strong perceptions that events rather than producers' actions
have led to the crisis in agriculture. More than 90 percent noted high
interest rates and low prices, and more than 80 percent noted high
costs for equipment and supplies. In addition, more than 70 percent
indicated changing land values and changing export markets were the
major causes for the crisis. Less than 30 percent noted technology, poor
marketing and management, and excessive living standards as causes
of the crisis. The pattern of greater emphasis on circumstances than on
actions is thus even more apparent among former producers than among
current producers experiencing financial stress.

The bottom panel of Table 6.5 suggests that former producers are
also the most favorable toward federal and state assistance with nearly
60 percent agreeing or strongly agreeing with the need for federal
assistance and 44 percent supporting state assistance (compared to only
26.9 percent of current producers with no debt supporting federal aid
and only 18.5 percent in the no debt category supporting state assistance).
In addition, of those agreeing or strongly agreeing with the need for
federal or state aid, more than 70 percent were in favor of each of the
4 specific types of aid being offered by the state and the federal
government. This group was particularly in favor of assistance for training
producers for new occupations.

Overall, former producers' responses show the patterns that might
be expected of persons who have been severely impacted by the farm

Table 6.5

**Responses of North Dakota Former Farmers to Selected Questions Concerning
the Reasons for the Farm Crisis and Agreement With Federal and State
Assistance for Financially Troubled Producers**

Selected Questions	Important	Somewhat Important
Perceived Reasons for the Crisis	(- - -percent- - -)	
Low farm yields	33.1	36.7
Farmers' making poor marketing decisions	25.6	44.6
Farmers' not using up-to-date technology	16.2	39.5
Bad weather	45.2	41.1
High interest rates	94.7	4.7
Low prices for farm products	95.2	4.2
Government involvement in agriculture	51.2	39.0
Corporate farms	16.8	29.9
Farmers' attempting to expand the size of their farms too rapidly	53.0	34.5
Farmers' being poor managers	31.0	41.7
The high cost of farm supplies/equipment	82.7	16.1
Changing land values	74.4	23.8
Changing export markets for farm products	73.2	23.8
Farmers' living beyond their means	25.7	47.3

Reason	Percent Listing the Reason as the Most Important
Low farm prices	43.4
High interest rates	22.6
Government involvement	8.2
High cost of supplies and equipment	4.4

Selected Questions	Strongly Agree	Agree	Neither Agree or Disagree	Disagree	Strongly Disagree
Financial Assistance for Producers	(- - - - - - - -percent- - - - - - - - -)				
The Federal government should assist farmers who are in financial trouble	19.0	39.9	18.4	21.5	1.2
The state government should assist farmers who are in financial trouble	13.8	31.4	23.9	28.3	2.5

Table 6.5 (Continued)

Selected Questions	Agree	Disagree
Type of Federal Aid[1]		
Providing financial assistance to financially troubled ag creditors either directly or through loan	(- -percent- -)	
guarantees	72.4	27.6
Subsidizing interest rates on operating loans	86.2	13.8
Participating with creditors and farmers in restructuring land debt	89.1	10.9
Providing low-interest loans or grants to financially stressed farm families for vocational training or college in preparation for a new occupation	98.9	1.1
Type of State Aid[1]		
Providing financial assistance to financially troubled ag creditors either directly or through loan guarantees	76.5	23.5
Subsidizing interest rates on operating loans	91.4	8.6
Participating with creditors and farmers in restructuring land debt	95.7	4.3
Providing low-interest loans or grants to financially stressed farm families for vocational training or college in preparation for a new occupation	98.6	1.4

[1]Noted by those agreeing or strongly agreeing with the need for aid.

crisis. They perceive circumstances that were largely outside their control as being the major causes of the crisis and strongly support the need for assistance to displaced farmers. Their responses tend to be a further accentuation of the responses of current producers with high debt-to-asset ratios.

Table 6.6 presents evaluations by current business operators, former business operators and other community residents regarding the farm crisis. The data in this table generally show business operators (both current and former) and community residents to have perspectives that are very similar to those of producers regarding agriculture in general and in terms of most of the reasons for the crisis (e.g., nearly 70 percent in each group cite high interest rates, 80 percent cite low prices, and over 50 percent cite export markets as very important causes for the crisis). They are somewhat more likely than producers to note reasons related to producers' actions as major causes of the crisis, however. Higher levels of agreement with reasons such as farmers' being poor managers (rated very important by roughly 40 percent in each of the three groups), farmers' living beyond their means (nearly 50 percent in all three groups rated this as very important), and similar responses are evident among each of the community groups. Finally, it is evident that levels of support for federal and state aid among business operators are similar to the levels of support among producers and that, among community residents, support is higher than among several categories (in terms of debts relative to assets) of producers. Thus, more than 50 percent of residents supported federal aid, and 46 percent supported state aid compared to 27 percent of producers with no debt who supported federal aid and 18 percent who supported state aid. Overall, although they tend to place a somewhat greater share of the responsibility for the crisis on producers, business operators (both current and former) and other residents tend to perceive the general conditions in agriculture, the causes of the crisis, and the need for federal and state support similarly to producers and, in fact, may be somewhat more sympathetic to the need for aid than producers with little or no debt.

Impacts on Producers, Other Rural Residents, and Rural Communities

Having examined producers' and residents' perceptions of farming and of the causes of the farm crisis, the next issue is the central one of what are the impacts of the crisis on producers and other residents in rural areas. Although data on such objective indicators of impacts as dollars of income lost and reduced business volume are difficult to obtain, information on perceived impacts can be obtained by examining

Table 6.6

Responses of Current and Former Business Operators and Community Residents to Selected Questions Concerning the General Conditions of Farming, the Causes of the Farm Crisis, and Assistance for Financially Troubled Farmers

Selected Questions	Chi Square	Current Business Operators	Former Business Operators	Community Residents
General Conditions	(− − − − − −percent agreeing− − − − − −)			
The proportion of farmers who are now in financial trouble is much greater than past	51.0	87.3	96.2	88.8
Most farms today are too large	146.8	70.1	97.9	83.4
Agriculture is our nation's most basic industry	91.9	92.1	94.3	93.2
To me, farming is strictly a business	135.9	70.6	77.4	61.9
The family farm is rapidly going out of existence	52.9	73.6	75.5	81.5
Agriculture plays a vital role in the nation's economy	43.8	96.9	96.2	97.1
American farmers will always be able to produce enough food to feed America	103.9	75.0	85.7	66.8
Today, large corporations, not farmers, control agriculture	129.8	61.8	59.6	68.9
Farmers are primarily to blame for high food prices	223.1	4.8	2.0	5.4
Farmers who adopt new technology and production methods will survive	19.7	63.7	69.4	66.7

Selected Questions	Chi Square	Current Business Operators		Former Business Operators		Community Residents	
		Very Important	Somewhat Important	Very Important	Somewhat Important	Very Important	Somewhat Important
Perceived Reasons for the Crisis	(− − − − − − − − − − − percent − − − − − − − − − − − −)						
High interest rates	17.5	70.9	27.4	84.0	13.6	67.4	28.2
Low prices for farm products	32.5	87.1	12.3	96.3	3.7	78.5	19.6
Government involvement in agriculture	6.8	49.8	43.8	58.2	35.4	45.6	46.1
Corporate farms	12.8	20.7	42.1	22.4	36.8	24.9	45.6
Farmers' attempting to expand the size of their farms too rapidly	27.9	60.3	32.5	63.0	30.9	50.8	34.7
Farmers' being poor managers	36.0	44.5	47.7	45.7	33.3	39.3	43.5
The high cost of farm supplies and equipment	9.8	64.9	33.1	68.3	27.3	69.8	26.7
Changing land values	26.5	57.9	36.9	66.7	24.7	50.4	38.2
Changing export markets for farm products	35.6	56.7	41.5	68.4	27.9	54.0	37.8
Farmers' living beyond their means	28.7	50.3	39.8	44.4	27.2	53.7	32.9

Table 6.6 (Continued)

Selected Questions	Chi Square	Current Business Operators	Former Business Operators	Community Residents
Financial Assistance for Producers		(percent agreeing/strongly agreeing)		
Agreement With Need for Federal Aid				
(1) Strongly agree	123.8	8.0	4.4	5.1
(2) Agree		26.2	31.1	46.2
(3) Neither agree nor disagree		23.9	27.8	16.3
(4) Disagree		27.3	26.4	29.0
(5) Strongly disagree		15.5	8.9	2.3
Types of Federal Assistance[1]				
Providing financial assistance to financially troubled ag creditors either directly or through loan guarantees	3.7[a]	61.1	64.3	55.8
Subsidizing interest rates on operating loans	44.8	64.9	60.0	52.2
Participating with creditors and farmers in restructuring land debt	38.2	62.5	62.5	55.7
Providing low-interest loans or grants to financially stressed farm families for vocational training or college in preparation for a new occupation	0.6[a]	55.7	62.5	46.2
Agreement With Need for State Aid				
(1) Strongly agree	143.8	6.1	2.1	2.6
(2) Agree		21.2	31.2	43.7
(3) Neither agree nor disagree		23.0	20.8	15.4
(4) Disagree		33.8	37.5	35.7
(5) Strongly disagree		15.9	83.3	2.6
Types of State Assistance[1]				
Providing financial assistance to financially troubled ag creditors either directly or through loan guarantees	47.1	69.5	76.9	56.6
Subsidizing interest rates on operating loans	9.7	69.2	76.9	55.9
Participating with creditors and farmers in restructuring land debt	9.0	71.0	85.7	59.9
Providing low-interest loans or grants to financially stressed farm families for vocational training or college in preparation for a new occupation	4.7[a]	57.7	85.7	56.4

Note: Chi squares were calculated from contingency tables of responses for each variable by the three respondent types. Percentages were calculated within respondent types.
[a]Chi square statistic is **not** significant at .05 level.
[1]Noted by those agreeing or strongly agreeing with the need for aid.

Table 6.7

Responses of Current Agricultural Producers to Questions Concerning the Impacts of the Crisis on Their Personal Lives and on Their Communities by Producers' Debt-to-Asset Ratios

Selected Questions	Debt-to-Asset Ratio				Total Sample
	No Debt	.01-.40	.41-.70	> .70	
Personal Impacts	(- - - - - - - - percent - - - - - - - -)				
Extent of Impact					
A great deal	15.6	28.0	46.1	55.8	32.1
Some	53.5	54.7	50.2	34.9	51.3
Not at all	30.9	17.3	3.7	9.3	16.7
$(x^2 = 47.2)*$					
	(- - - - percent experiencing impact - - - -)				
Types of Personal Impacts					
Lost a farm due to financial difficulties	1.7	3.6	6.0	9.3	4.3
Lost a business due to financial difficulties	1.7	3.7	5.6	7.0	4.0
Lost a job because a business had to cut back its staff	12.3	14.7	15.7	11.6	14.0
Had a reduction in pay, benefits, or working hours because a business had to cut back	22.9	18.7	21.0	23.4	20.6
Lost a home, car or other major possession to a finance company or bank	1.7	2.2	5.6	6.4	3.2
Suffered depression or other emotional problems	17.6	23.1	34.8	43.0	26.6
Committed suicide	0.7	0.7	0.4	1.2	0.7
Experienced unusual marital or other family stress or conflict	9.7	14.7	22.1	26.7	16.5
Been divorced or separated	3.0	6.3	5.6	9.9	5.9
Been convicted of a crime other than a minor traffic violation	0.3	0.4	0.4	0.6	0.4
Community Impacts	(- - - - - - - - - percent - - - - - - -)				
Extent of Impacts					
A great deal	34.1	49.7	62.2	78.6	49.6
Some	47.5	40.9	32.2	17.9	39.6
Not at all	18.4	9.4	5.6	3.6	10.8
$(x^2 = 155.4)*$					
	(- - - percent experiencing impacts - - - -)				
Types of Community Impacts					
Employment opportunities for farm workers	63.7	75.4	92.1	85.7	75.4
Employment opportunities for nonfarm workers	55.7	62.8	75.9	81.8	64.2
Number of agriculturally related businesses (agri-business firms--not farms)	57.1	71.9	82.2	89.3	70.9
Employment in agiculturally related businesses	57.7	75.0	83.3	88.9	72.7
The number of nonagricultural businesses	50.0	59.6	67.4	63.0	58.4
Employment in nonagricultural businesses	50.6	62.9	70.8	67.9	61.1
Public schools	26.7	35.2	40.7	55.4	35.4
Law enforcement	12.4	11.3	9.9	22.2	12.3
Medical services	12.5	17.8	18.7	29.6	17.5
Mental health care	1.8	6.9	6.7	11.5	5.9
Churches	24.7	33.1	35.6	46.4	32.3
Housing	37.6	51.9	51.1	58.9	48.6
Property taxes	50.0	53.7	54.9	56.4	53.1
Utility rates	33.0	34.9	23.1	40.7	33.2
Cost of living	52.0	49.1	50.5	51.9	50.3
Crime rates	36.9	42.9	37.1	33.9	39.8
Fire protection	4.5	6.6	4.4	9.3	6.0
Community spirit or morale	39.4	58.9	59.3	89.3	56.3

*Significant at the .05 level.

the responses of producers and other residents in rural areas to questions concerning the impacts on their personal lives and on their communities.

Table 6.7 presents the responses of producers to two major questions related to impacts. As shown in this table, these questions asked what effects the current crisis has had on respondents' personal lives and on their communities. An examination of these data for current producers (see the top panel) suggests that the crisis has had a significant effect on both their personal lives and their communities. More than 83 percent of the respondents believe that the crisis has had a great deal or at least some effect on their personal lives, and nearly 90 percent believe it has had some or a great deal of an effect on their communities. Although a larger proportion of producers with higher levels of debt believe that they have been affected (as might be expected), even among those with no debt, a large majority indicate that they and their communities have been affected by the crisis.

Among the personal effects noted most often are depression and other emotional problems, reduction in pay or other benefits, and unusual marital or family stress or conflict. These effects are especially prevalent among those with the highest levels of debt relative to assets (i.e., more than 70 percent). Among this group, 43 percent had experienced depression, nearly 27 percent had experienced marital or family stress, and 23 percent had had reductions in pay or benefits. In terms of community effects, the most often noted effects were those on employment, businesses, housing, and taxes. Although the extent to which a particular effect was noted by respondents varied somewhat by their debt-to-asset ratio, variation among categories of debts to assets was less than for personal effects. Overall, the data in Table 6.7 suggest that agricultural producers perceive that they and their communities have been impacted in largely negative ways by the current farm crisis.

Table 6.8 presents information on the personal effects noted by the sample of North Dakota former farmers. Questions related to community impacts were not asked because former farmers had often relocated by the time of the interview. An examination of the data in Table 6.8 indicates that 97 percent of these respondents indicated that their lives had been affected somewhat or a great deal by the farm crisis. Of the effects noted, those listed most often included depression (48.5 percent), marital problems (42.0 percent), reduction in pay or benefits (22.5 percent), and job loss (21.9 percent). These findings are clearly as expected for a group of persons who have suffered serious economic hardships because of the crisis but nevertheless substantiate the social and emotional costs associated with the crisis for those directly impacted.

Table 6.9 presents data on the impacts perceived by current and former business operators and other residents of rural communities. The

Table 6.8

Responses of North Dakota Former Farmers to Questions Concerning the Impacts of the Crisis on Their Personal Lives

Selected Questions	Percent
Extent of Personal Impact	
A great deal	68.6
Some effect	28.4
Not at all	3.0
Type of Personal Impact*	
Lost a farm due to financial difficulties	55.0
Lost a business due to financial difficulties	11.8
Lost a job because a business had to cut back staff	21.9
Had a reduction in pay, benefits or working hours because a business had to cut back	22.5
Lost a home, car, or other major possession to a finance company or bank	18.3
Suffered depression or other emotional problems	48.5
Committed suicide	1.2
Experienced unusual marital or other family stress or conflict	42.0
Been divorced	13.6
Been convicted of a crime other than a minor traffic violation	3.5
None	10.1

*Percent experiencing impact.

Table 6.9

**Responses of Current and Former Business Operators and Other Rural
Residents to Questions Concerning the Impacts of the Crisis on Their
Personal Lives and on Their Communities**

Impacts	Current Business Operators	Former Business Operators	Other Community Residents
Personal Impacts	(– – – – – –percent– – – – – –)		
Extent of Impacts			
A great deal	24.4	51.9	18.0
Some	51.7	40.7	44.8
Not at all	23.9	7.4	37.2
(x^2 = 84.2a)	(–percent experiencing impacts–)		
Type of Impacts			
Loss of business	7.1	23.9	9.1
Loss of job	10.4	20.7	21.0
Reduced pay	13.2	13.2	27.1
Loss of home/major possession	4.5	21.7	7.7
Depression	11.5	11.9	16.4
Suicide	1.8	0.0	1.3
Family stress	10.2	13.5	12.2
Divorce	6.8	8.3	9.8
Convicted of a crime	0.6	0.0	2.6
Community Impacts	(– – – – – – percent – – – – – –)		
Extent of Impacts			
A great deal	77.0	---	63.7
Some	22.3	---	32.0
Not at all	0.7	---	4.3
(x^2 = 41.3[a])	(– – percent noting impacts – –)		
Types of Impacts			
Farmers going out of business	32.6	---	34.6
Other businesses going bankrupt	11.8	---	10.4
Bad economy	14.4	---	13.0
No/few jobs/not hiring	4.2	---	5.9
Emotional stress	7.4	---	3.9
Many houses for sale/not selling	1.1	---	1.2
Layoffs/people losing jobs	34.2	---	21.5
People moving to cities for jobs	8.3	---	6.5
No money/cash flow down	13.9	---	14.6
Spending down	6.7	---	6.3
Loss of machinery	2.8	---	2.3
Banks closing	10.6	---	8.2
Wife has to work	0.0	---	0.0
Other	12.9	---	17.3
N =	(714)	(83)	(829)

Table 6.9 (Continued)

Impacts on Services	Former Business Operators			Community Residents		
	Great Deal	Some	Not at All	Great Deal	Some	Not at All
Services	(- - - - - -percent noting impact- - - - - -)					
Schools	22.1	51.0	26.9	17.1	45.6	37.3
Law enforcement	3.6	31.3	65.1	1.1	11.9	87.1
Medical services	7.7	35.9	56.4	4.4	26.2	69.4
Mental health care	7.3	32.5	45.9	5.5	13.5	81.0
Churches	10.7	52.7	36.6	10.1	40.9	49.0
Public welfare	19.3	45.6	35.2	11.4	36.3	52.3
Family counseling and other services	12.7	44.3	43.1	7.1	19.2	73.8
Fire protection	2.1	19.5	78.5	0.4	4.3	95.4
Other municipal services	3.3	31.9	64.8	0.9	7.5	91.7

[a]Significant at the .05 level.

data in this table generally show that persons in rural communities perceive that their community has been impacted by the crisis. Thus, 76 percent of the current business operators, 93 percent of the former business operators, and 63 percent of the community residents believe that they personally have been affected a great deal or somewhat by the crisis, and the percentages noting some or a great deal of effect on their community are 99 percent for current business operators and 96 percent for community residents (former business operators were not asked questions related to the community).

The effects on personal lives that were identified most often include reduced income (13 percent), depression (11.5 percent), job loss (10 percent), and family stress (10 percent) for current operators; loss of major possessions (22 percent), job loss (21 percent), family stress (14 percent) and reduced income (13 percent) for former business operators (and of course loss of business); and reduced income (27 percent), job loss (21 percent) and depression (16 percent) for community residents. Overall, the percentage indicating the occurrence of these events is substantially lower than for producers. A similar pattern is evident for effects on communities. Of the most often noted impacts, the loss of farm operators and job loss are the most often noted impacts by both current business operators (33 percent cited the loss of producers and 34 percent job loss) and community residents (35 percent and 22 percent). It is significant, however, that community services are generally seen as not having been significantly impacted by the crisis, particularly by community residents.

Overall, the data on personal and community impacts suggest that agricultural producers, rural business operators and rural nonfarm, nonbusiness community residents perceive that they and their communities have been impacted by the farm crisis. Among the most noted impacts are the loss of producers, the loss of jobs for other rural residents, increased incidences of depression, marital and family conflicts and similar personal problems. There is substantial agreement among these different groups in the ranking of the most important impacts relative to other effects, but variation among groups in their perceptions of the frequency of impacts is substantial. The degree to which impacts are perceived, both personally and for their community, appears to be closely related to the degree of direct impacts experienced. Producers, and particularly those who have been forced to leave agriculture, perceive not only more personal effects but also more community impacts than other rural residents, and within rural communities, business operators are more likely to note that impacts have occurred than other residents. The extent to which respondents have been directly affected by the crisis clearly affects their perceptions of impacts.

Summary and Conclusions

The analysis reported in this chapter shows that those being most directly impacted by the crisis are producers and business operators in early career stages who entered agriculture or business in the 1970s. They are as well educated and innovative as those who have been able to remain in agriculture and business but initiated their farms and firms during periods of high prices and high costs and were not sufficiently established to meet debt obligations when prices and business volumes declined. Producers perceive that macroeconomic factors, such as high interest rates, low commodity prices, reduced exports and similar factors have brought about the crisis, but they believe that agriculture remains at the center of the United States' economy. Producers, business operators, and other community residents believe that they and their communities have been negatively impacted by the farm crisis with the economic characteristics of the employment base and of business activity being most negatively impacted, but with personal impacts such as increased levels of depression and other emotional problems as well as marital and other family problems also being apparent.

The results suggest that the impacts of the crisis, although varying in their intensity among different groups, are pervasive in rural society. There is thus a rural crisis, not just a farm crisis. The implications of the loss of young innovative producers and business persons in rural communities and the loss of income and job opportunities from rural areas, which in many cases already lacked sufficient job opportunities prior to the crisis, can only be seen as problematic for rural areas. Although the implications of such impacts will be more fully explored in the chapter to follow, the findings reported in this chapter reveal that the impacts of the farm crisis are apparent to the residents of rural areas and are impacting the structure of short-term and long-term opportunities in rural areas.

7

The Implications of the Current Farm Crisis for Rural America

Steve H. Murdock, Lloyd B. Potter, Rita R. Hamm,
Kenneth Backman, Don E. Albrecht
and F. Larry Leistritz

In the preceding chapters of this work we have presented an overview of the financial, demographic, socioeconomic, and service context in which the current crisis is occurring and examined the impacts of the crisis on producers and other residents of rural areas. In this final chapter, we examine the long-term implications of the current farm crisis for agriculturally dependent rural areas in the United States. Specifically, we examine the implications of the loss of alternative numbers of residents for these areas. Although the loss of a substantial proportion of producers will impact nearly all socioeconomic dimensions of rural areas, we restrict this examination, due to space and data limitations, to examining the potential quantitative effects on the structure of agriculture; on the economic, demographic, public service, and fiscal characteristics of agriculturally dependent rural areas; and to examining the potential qualitative effects of changes in these dimensions on the social structure of rural areas.

In presenting the analysis reported here, it has been necessary to make a number of very speculative assumptions. We have made assumptions about the proportion of producers who will leave farming and rural areas in the coming years, about the proportion of sales that will be lost to rural communities due to the loss of these producers, about the numbers of persons (both producers and related secondary worker populations) who will migrate from rural areas due to the crisis, and about the public service and fiscal implications of the loss of alternative numbers of persons from rural areas. For each of these factors, we have been forced to make assumptions that extend beyond the base of existing knowledge. Thus, there is a limited empirical base of evidence from which to estimate the exact proportion of producers with different levels of debt who will actually fail in the coming years, about the time period over which those producers who fail will leave farming, or about the local area impacts of such a loss. It is clear, for example, that some producers with even the highest levels of debt may be able to restructure that debt and remain in farming. In addition, in nearly all cases, the land farmed by producers who fail will remain in production (albeit under the direction of new operators). As a result, many of the expenditures associated with production activities (e.g., purchases of fertilizers, seed) will continue to be made even if they are made by a smaller number of producers. The economic impacts of the loss of a given number of producers cannot therefore be easily estimated by the use of a simple export-base multiplier.

In addition, it must be recognized that the loss of sales within one area is often offset by a gain in sales in another area. Sales lost to small rural towns and businesses may be gained by larger cities in areas quite distant from a rural area. Because the number of producers who may be able to remain in rural areas in new nonfarm-related jobs is uncertain, the retail sales, service requirements, and fiscal impacts of the crisis cannot be determined with certainty. However, although such implications cannot be measured with certainty, agriculturally dependent rural areas have continued to lose population and business establishments during the past 60 years, despite a continued increase in agricultural production and relatively little change in cultivated acreage. This continued decline in agriculturally dependent rural areas provides evidence that such implications are real, even if they are difficult to assess.

Because of the uncertainty surrounding the estimation of secondary impacts, we have made two alternative assumptions about this factor and have restricted our focus to the likely impacts of farm failure on agriculturally dependent counties and trading centers in the United States. We fully realize that for the United States as a whole, for overall levels of agricultural production, and for consumers and other groups, the implications may be quite different than those for persons in agriculturally dependent rural areas. For these larger areas, the implications may be largely neutral or, in fact, positive. As noted in the introduction, then, we admit to a bias in that we are most concerned in this analysis about the implications of the crisis for rural areas in the United States.

The intent of the analysis is thus to examine the possible long-term implications of the farm crisis on agriculturally dependent rural areas. *We are under no illusion that the implications estimated will reflect the actual course of events that will result from the crisis. Rather, it is because of the uncertainty regarding such implications that we believe that an examination of a range of potential implications of the crisis, even if based on rather speculative assumptions, is essential for discerning the types and magnitude of policy actions that should be formulated to address such implications.* In the absence of attempts to quantitatively estimate these implications, policy formation is likely to remain largely rhetorical.

Methodology for Estimating Implications

As noted above, the analysis reported in this chapter makes a number of assumptions that must be described before the results of the analysis are reported. First, the analysis is restricted to the potential implications of farm failure for agriculturally dependent counties. Following the lead of Bender et al. (1985), we define agriculturally dependent counties as those nonmetropolitan counties in which 20 percent or more of total earnings is from agriculture. However, whereas Bender et al. utilized a weighted average of income and employment from 1975–1979 and the designation of metropolitan areas as delineated in 1974, we utilize average earnings from agriculture in 1976, 1980, and 1983 and the metropolitan designation used for the 1980 Census. We made these alterations of the procedures utilized by Bender et al. because we deemed it desirable, because the farm crisis had its origins in the events of the late 1970s but is having its most obvious impacts in the mid-1980s, to utilize data that would span the period of time from the late 1970s to the beginning of the 1980s. The procedures we utilize provide a set of counties with agricultural involvement throughout the major period of the evolution of the causes and consequences of the crisis. The period chosen for measuring agricultural dependence is important because of the rapid decline that has occurred in the number of agriculturally dependent counties. If the 1980 delineation of metropolitan status and the 20 percent criteria are used, 639 counties can be identified as agriculturally dependent in 1976, but by 1983 this number is only 362. There has thus been a rapid decline in agriculture as an income source in the United States during the early 1980s. We utilized the metropolitan designation in 1980 because a majority of the other data used in the analysis presented below are from the 1980 Census. These procedures resulted in 472 nonmetropolitan counties being identified as agriculturally dependent. Throughout the analysis presented below, we contrast events in these agriculturally dependent counties to those in other nonmetropolitan counties, in metropolitan counties, and in all U.S. counties.

The extent to which the counties so designated show other characteristics expected in agricultural counties can be determined by examining the data in Tables 7.1 and 7.2. As shown in the data in Table 7.1, the

Table 7.1

**Selected Demographic Characteristics of Agriculturally Dependent
Nonmetropolitan Counties, Other Nonmetropolitan Counties, Metropolitan
Counties, and All U.S. Counties, 1980**

	County Type			
Selected Characteristics	Agriculturally Dependent Counties	Other Nonmetro Counties	Metropolitan Counties	All U.S. Counties
Total population	4,081,088	53,034,094	169,430,623	226,545,805
Percent				
Urban	21.7	38.1	85.3	73.7
Rural	78.3	61.9	14.7	26.3
Rural farm	19.2	6.6	1.0	2.5
< 18 years of age	30.2	29.4	27.7	28.1
18-24 years of age	10.9	12.7	13.5	13.3
25-34 years of age	13.6	14.9	16.9	16.4
35-44 years of age	10.2	10.7	11.5	11.3
45-54 years of age	9.9	10.7	10.2	10.1
55-64 years of age	10.4	7.3	9.5	9.6
65+ years of age	14.9	12.9	10.7	11.2
Black	6.7	9.0	12.6	11.7
Spanish origin	6.9	2.8	7.5	6.5
Number of households	1,431,238	17,357,039	61,601,325	80,389,602
Average size of household	2.9	2.9	2.8	2.8
Percent of households that are family households	76.8	76.4	72.3	73.3
Percent married-couple households with children present < 18 years of age	35.0	34.5	30.6	31.5

Table 7.2

**Selected Agricultural Characteristics of Agriculturally Dependent
Nonmetropolitan Counties, Other Nonmetropolitan Counties, Metropolitan
Counties, and All U.S. Counties, 1982**

	County Type			
Selected Characteristics	Agriculturally Dependent Counties	Other Nonmetro Counties	Metropolitan Counties	All U.S. Counties
Number of farms	284,341	1,308,359	648,276	2,240,976
Percent of farms with sales:				
< $10,000	28.5	49.8	56.4	49.0
$10,000-$19,999	12.1	11.9	10.7	11.6
$20,000-$39,999	15.6	11.1	9.1	11.1
$40,000+	43.8	27.2	23.8	28.3
$100,000+	19.9	12.8	12.2	13.5
Farmland as a percent of total land	77.4	37.4	39.2	43.5
Percent of farmland acres irrigated	25.1	25.9	21.4	25.0
Percent of harvested cropland	32.7	31.1	39.8	32.9
Average size of farm (acres)	889	437	249	440
Average value of farm land and buildings	$483,514	$312,383	$352,841	$345,801
Percent working 100+ days off-farm	31.8	47.1	53.3	47.0

counties designated as agriculturally dependent have only 21 percent of their populations living in urban areas of 2,500 or more compared to 74 percent in the nation as a whole, while 19 percent of their residents live on farms compared to only 2.5 percent in the nation as a whole. In addition, they have the relatively high proportions of children and older adults and larger proportion of family-oriented households typical of more traditional rural areas.

The agricultural base of these counties is particularly evident when the data in Table 7.2 are examined. Nearly 78 percent of all land in these counties is in farms compared to only 44 percent for the nation as a whole, and the average size of farm was more than 880 acres, compared to 440 acres for the nation as a whole. In addition, although these counties contain only 10 percent of all farms, they have largely commercial-scale farms (i.e., farms with sales of $40,000 or more), and only 28 percent of their farms had sales of less than $10,000. It is evident that the delineation procedure utilized has isolated areas that have economies that are largely concentrated in agriculture and have other characteristics of rural areas.

Perhaps the most critical assumption in terms of the analysis in this chapter, however, is that related to the number of producers who will be forced to leave agriculture. Although any such estimate must clearly be speculative, existing analyses (Dunn 1987; Ginder et al. 1985; Goreham et al. 1987; Marousek 1979; Barry 1986) suggest that a majority of those who have debt-to-asset ratios of more than 70 percent will be unable to remain in farming and that a significant percentage of those in the 41 to 70 percent debt-to-asset ratio category may also be forced to leave agriculture. Using the estimates for 1985 of the number of farms by debt-to-asset ratio of Johnson et al. (1986), we thus assume that all of those who have debt-to-asset ratios exceeding 100 percent will be forced from farming (such producers are in fact already insolvent), that 75 percent of those with debt-to-asset ratios of 70 to 100 percent will fail, and that 50 percent of those with debt-to-asset ratios of 41 to 70 percent will fail. Although admittedly simplistic, given that the estimates by Johnson et al. (1986) do not address failure among smaller U.S. farms because these are largely omitted from their survey efforts, we believe that our estimates of the number of farms likely to fail are probably conservative.

Using these rates of failure, we estimate that by about 1995, 61,000 producers who had debt-to-asset ratios of more than 100 percent in 1985 will fail, along with 54,000 producers who had debt-to-asset ratios of 71 to 100 percent and 98,500 producers with debt-to-asset ratios of 41 to 70 percent. We thus estimate that 213,500 producers will discontinue farming in agriculturally dependent counties by 1995. Admittedly not all of the producers who will fail will live in agriculturally dependent counties, but we believe a majority will be from these areas and thus for illustrative purposes we make the simple assumption that these producers will be from agriculturally dependent counties. We make the additional simplistic assumption that, because these producers are located primarily in counties with few alternatives to agriculture, they will also be forced to relocate outside of agriculturally dependent counties.

. An estimation of the secondary effects of the loss of a given number of producers is equally difficult to complete because it is evident that many of the secondary impacts of expenditures in agriculture occur outside rural farming areas. As a result, we have made relatively conservative assumptions about the secondary effects of the failure of producers. Given that total multipliers for agriculture tend to vary from 2.5 to nearly 4.0 when the location of such impacts is not considered and given that most analysis of the secondary impacts of resource-based developments in rural areas (see Murdock et al. 1986b) show that the multiplier effects in local rural areas are often less than 2.0 (including the direct worker), we have made 2 alternative assumptions concerning secondary employment loss resulting from the loss of producers in agriculturally dependent rural areas. These assumptions attempt to simultaneously take into account that our interest is in secondary impacts only within agriculturally dependent counties themselves and in those secondary workers who will be forced to leave these counties as a result of the crisis. The alternative assumptions used are that for each producer lost, either .25 secondary workers (scenario I) or .75 secondary workers (scenario II) will be lost from agriculturally dependent counties. The use of these assumptions results in the estimation that between 53,000 and 160,000 secondary workers could be lost from agriculturally dependent rural areas as a result of farm failure in these areas.

Given assumptions about the number of producers and secondary workers who will leave agriculturally dependent nonmetropolitan counties as a result of the crisis, the remainder of the analysis was completed by examining the implications of the loss of producers and secondary workers on the structure of agriculture and on the economic, demographic, public service, fiscal, and social structure of agriculturally dependent counties. This was done by using data on the agricultural, economic, demographic, public service, fiscal, and social characteristics of such counties derived from the 1980 Census of Population and Housing and the 1982 Censuses of Agriculture, Government, and Business; by utilizing data from the North Dakota and Texas surveys of producers, former producers, current and former business operators, and community residents; and by using data from the U.S. Department of Agriculture on the characteristics of producers with alternative levels of debt in 1985

(Johnson et al. 1986.) Thus we assume that those producers who will leave agriculturally dependent counties will have the average characteristics of producers with high debt-to-asset ratios and of producers who have already been forced to leave farming during the mid-1980s. Average characteristics of former producers and of those in our surveys with debt-to-asset ratios exceeding 40 percent are used to measure the likely demographic characteristics of producers who will leave agriculture in the coming years.

Estimating the characteristics of secondary workers posed other problems. Because of the relatively small sample size in our former business surveys (only 83) and the fact that many who will leave will likely not be business operators but other community residents, we used the average characteristics of all respondents in our community surveys (former and current business operators as well as other community residents) as assumed characteristics of secondary workers who would leave agriculturally dependent counties in the years to come.

To assess the impacts on agriculture, we used data from Johnson et al. (1986) on the characteristics of producers with high debt-to-asset ratios (more than 40 percent) in 1985 and subtracted the estimated number of producers with specific characteristics from the total number of producers with such characteristics in agriculturally dependent counties as indicated in the 1982 Census of Agriculture. To assess the demographic impacts of the loss, we estimated population loss by applying our assumptions about the demographic characteristics of those who will be forced to leave rural areas (such as their average household size, number of dependents under 18 years of age, and the age distribution of the producers and their families) and applied these to the number of producers and secondary workers assumed to be leaving agriculturally dependent rural areas. These projected total population effects were then examined in relation to 1980 Census data for such counties. Because of the relatively young age distribution of the producer and secondary-worker populations being examined, we did not attempt to survive or otherwise extend the demographic groups for 1980 to 1995 before making comparisons. Rather impacts were examined relative to the characteristics of population in agriculturally dependent counties in 1980.

Economic, public service, and fiscal impacts were projected largely on a per population unit basis given the number of persons projected to leave rural areas. The per capita and population-unit-based rates (such as rates per 1,000 or per 100,000 persons) are presented in Tables 7.3 and 7.4. A summary of the key assumptions and methods used to project the number of persons assumed to leave agriculturally dependent rural areas, as well as a summary of the projected number and age characteristics of these populations, is presented in Table 7.5.

Table 7.3

Selected Economic and Business Characteristics of Agriculturally Dependent Nonmetropolitan Counties, Other Nonmetropolitan Counties, Metropolitan Counties, and All U.S. Counties

Characteristic	County Type			
	Agriculturally Dependent Counties	Other Nonmetro Counties	Metropolitan Counties	All U.S. Counties
Manufacturing/Business				
Bank deposits per capita, 1980	$5,199	$4,105	$6,044	$5,602
Bank savings per capita, 1980	$816	$831	$1,207	$1,117
Savings and loan savings per capita, 1979	$842	$1,223	$2,243	$1,993
Retail sales per capita, 1982	$2,977	$3,888	$4,982	$4,705
Manufacturing payroll per capita, 1979	$344	$1,049	$1,616	$1,468
Value added in manufacturing per capita, 1980	$8,522	$25,945	$37,348	$34,336
Private business establishments per 1,000 population, 1980	19.38	18.74	20.34	19.97
Manufacturing establishments per 1,000 population, 1980	1.08	1.30	1.46	1.42
Labor Force				
Civilian labor force, 1980	1,834,215	22,179,928	82,919,410	106,933,553
Percent of population in the labor force	44.90	44.50	48.00	47.20
Unemployment rate, 1980	6.80	8.10	6.90	7.10
Labor force, 1984	1,902,562	23,092,274	88,500,239	113,495,075
Unemployment rate, 1984	8.50	9.20	7.10	7.50
Per Capita Earnings ($)				
Agriculture, forestry, and fisheries, 1980	59.73	22.50	21.01	22.03
Mining, 1980	72.34	256.16	77.39	116.61
Construction, 1980	199.39	300.99	437.50	403.19
Manufacturing, 1980	510.88	1,354.82	1,983.81	1,818.95
Transportation and utility, 1980	259.20	325.72	604.93	537.30
Wholesale trade, 1980	216.07	1991.33	541.66	460.51
Retail trade, 1980	400.27	482.87	729.22	669.12
Finance, insurance, and real estate, 1980	134.59	156.98	476.76	400.27
Services, 1980	398.07	605.04	1,389.84	1,199.39
Government, 1980	736.26	875.04	1,251.04	1,159.08
Government (Federal), 1980	0.12	0.15	0.31	0.27

Table 7.4

Selected Service and Fiscal Characteristics of Agriculturally Dependent Nonmetropolitan Counties, Other Nonmetropolitan Counties, Metropolitan Counties, and All U.S. Counties

	County Type			
Characteristic	Agriculturally Dependent Counties	Other Nonmetro Counties	Metropolitan Counties	All U.S. Counties
Crimes per 100,000 population, 1980	2,133.21	3,155.63	6,553.82	5,726.88
Police officers per 1,000 population, 1980	1.37	1.38	2.03	1.87
Public school students per capita, 1980	0.21	0.20	0.18	0.18
Median years of school completed for persons 25 +, 1980	12.30	12.20	12.40	12.30
Total Federal expenditures per capita, 1983	$2,125.32	$2,044.11	$3,057.97	$2,819.73
Local government general revenue per capita, 1982	$1,115.28	$979.22	$1,281.69	$1,212.44
Percent intergovernmental funds from the state, 1982	39.24	38.16	32.90	33.93
Percent intergovernmental transfer payments of all revenues, 1982	44.04	44.22	40.85	41.49
Total revenue (total taxes) per capita, 1982	$366.90	$304.80	$490.65	$447.72
Percent of taxes from property tax, 1982	30.53	26.17	28.46	28.09
Government expenditures per capita, 1982	$1,069.59	$948.77	$1,197.14	$1,140.44
Percent expenditures for education, 1982	48.94	49.01	40.71	42.36
Percent expenditures for health and hospital, 1982	10.45	10.78	7.47	8.13
Percent expenditures for public welfare, 1982	3.28	2.83	6.23	5.56
Percent expenditures for highways, 1982	12.03	8.02	4.73	5.45
Percent expenditures for police, 1982	3.55	3.78	5.73	5.34
Debt outstanding per capita, 1982	$673.45	$871.39	$1,187.63	$1,109.18

Table 7.4 (Continued)

Characteristic	County Type			
	Agriculturally Dependent Counties	Other Nonmetro Counties	Metropolitan Counties	All U.S. Counties
Full time government employees per 1,000 population, 1982	38.35	33.30	33.54	33.57
Per capita property value of locally assessed property, 1979	$7,693	$5,536	$6,089	$5,997
Physicians per 100,000 population, 1980	53.12	93.94	225.99	193.84
Hospital beds per 100,000 population, 1980	434.79	576.00	625.01	610.81
Nursing home beds per 1,000 persons 65+ years of age 1978	64.65	56.22	51.20	52.79
Percent of housing units vacant, 1980	10.75	10.01	6.38	7.27
Percent of housing units without complete plumbing, 1980	4.26	4.30	1.52	2.17
Percent owner occupied, 1980	74.40	73.16	61.74	64.43
Percent renter occupied, 1980	22.85	24.15	35.82	32.98
Percent of 1980 housing units built from 1970-80	23.8	28.7	25.6	26.2
Percent of 1980 housing units built 1939 or earlier 38.9	30.4	24.2	25.9	
Per capita money income, 1979	$5,610	$5,938	$7,727	$7,295
Personal income per capita, 1979	$7,812	$7,514	$10,138	$9,519
Percent of population receiving Social Security benefits, 1980	19.24	17.90	14.47	15.31
Percent of families receiving Aid to Families with Dependent Children, 1980	4.17	4.84	6.42	6.02
Births per 1,000 population, 1980	17.20	16.33	15.78	15.93
Deaths per 1,000 population, 1980	9.23	11.58	10.28	10.55

Table 7.4 (Continued)

	County Type			
Characteristic	Agriculturally Dependent Counties	Other Nonmetro Counties	Metropolitan Counties	All U.S. Counties
Infant deaths per 1,000 population, 1980	19.40	18.20	18.30	18.30
Marriages per 1,000 population, 1980	10.45	9.61	8.50	8.78
Divorces per 1,000 population, 1980	3.60	4.99	5.36	5.25
Percent of persons below poverty, 1979	17.75	15.06	11.10	12.09
Percent of persons 65+ below poverty, 1979	20.24	19.89	11.77	14.02
Percent of persons < 18 below poverty, 1979	21.77	18.36	14.77	15.73

Table 7.5

Summary of Assumptions
Used to Project Demographic Implications of the Farm Crisis
in Agriculturally Dependent Counties in the United States, 1985–1995,
by Step in the Estimation Process

Step 1: Assumed Percent and Number of Producers Leaving Agriculture and
Agriculturally Dependent Counties by Debt-to-Asset Ratio

Debt-to-Asset Ratio	Assumed Percentage Leaving Agriculture (1985–1995)	Number of Producers in the U.S. in 1985	Number Assumed to Leave Agriculture (1985–1995)	Percent of All Producers in the U.S. (1,551,000)[1]
> 1.00	100	61,000	61,000	3.9
.71-1.00	75	72,000	54,000	3.5
.41-.70	50	197,000	98,500	6.4
Total	----	330,000	213,500	13.8

Step 2. Assumed Percent and Number of Secondary Workers Leaving
Agriculturally Dependent Counties in the United States (1985–
1995), Under Two Scenarios

Scenarios	Assumptions	Secondary Workers Assumed to Leave Agriculturally Dependent Counties
I	.25 secondary workers leave area for every producer leaving the area	53,375
II	.75 secondary workers leave area for every producer leaving the area	160,125

Table 7.5 (Continued)

Step 3. Population Leaving Agriculturally Dependent Counties in the U.S.

a. Producer-Related Population

Debt-to-Asset Ratio	Producers Lost	Assumed Household Size[2]	Total Population Lost
1.00 +	61,000	3.8	374,300
.71–1.00	54,000	3.8	205,200
.41–.70	98,500	3.8	31,800
Total	213,500	----	811,300

b. Secondary-Worker-Related Population

Scenario	Secondary Workers Lost	Assumed Household Size	Worker-Related Population Lost
I	53,375	3.1	165,463
II	160,125	3.1	496,388

c. Total Population

Scenario	Total Producer-/ Secondary-Worker- Related Population Lost
I	976,763
II	1,307,688

Table 7.5 (Continued)

Step 4. **Assumed Number of Persons Less Than 18 Years of Age Leaving Agriculturally Dependent Counties**

a. Producer-Related Population

Producers Leaving	Assumed Persons per Household < 18	Population < 18 Leaving
213,500	1.4	298,900

b. Secondary-Worker-Related Population

Scenario	Secondary Workers Lost	Population < 18 Lost
I	53,375	74,725
II	160,125	224,175

c. Total Population

Scenario	Total Population < 18 Leaving Agriculturally Dependent Counties (Both Producer- and Secondary-Worker Related)
I	373,625
II	523,075

Table 7.5 (Continued)

Step 5. **Assumed Age Structure of Producers and Other Rural Residents**
(Secondary-Worker Populations) Leaving Agriculturally Dependent
Rural Areas[3]

	Producers and Other Rural Residents Leaving			
	Scenario I		Scenario II	
Ages[3]	(Number)	(Percent)	(Number)	(Percent)
< 18	373,625	38.2	523,075	40.0
18-34	215,729	22.1	271,593	20.8
35-44	165,616	17.0	216,193	16.5
45-54	122,947	12.6	162,184	12.4
55-64	74,479	7.6	110,275	8.4
65 +	24,367	2.5	24,367	1.9
Total	976,763	---	1,307,688	---

---

[1]Data for 1985 derived from Johnson et al. 1986.
[2]Data derived from surveys conducted by the authors in 1985 and 1986
(see chapter 6).
[3]Age distributions of producers by debt category were taken from
Johnson et al. 1985:48, ages of others were assumed to be
proportional to the population in agriculturally dependent counties
in 1980 (see Table 7.1).

Table 7.6

**Summary of Characteristics of the Estimated Number
of Producers Leaving Agriculture, 1985–1995**

Farms in Agriculturally Dependent Counties	Estimated Number of Producers Leaving Agriculture	Estimated Percentage of All Producers in Agriculturally Dependent Counties Leaving Agriculture
284,341	213,500	75.1*

Age Distribution of Producers Leaving Agriculture

Age	Percent
<34	33.1
35–44	30.8
45–54	22.2
55–64	11.2
65+	2.7

Distribution of Farms Estimated to Cease Operation by Crop Type

Farm by Crop Type	Estimated Percent Ceasing Operation
Cash grain	36.2
Tobacco and cotton	4.7
Vegetable, fruit/nut	3.6
Nursery or greenhouse	1.4
Other crop	4.9
Beef, hog and sheep	27.6
Dairy	16.9
Poultry and eggs	2.0
Other	2.7

Distribution of Farms Estimated to Cease Operation by Class of Sales

Farm by Class of Sale	Estimated Percent Ceasing Operation
$250,000 +	13.7
$100,000-$249,999	26.3
$40,000- $99,999	24.4
$20,000- $39,999	10.5
$10,000- $19,999	9.2
Less than $10,000	15.9

*This percentage of farm failure obviously will not occur. It is presented here only to illustrate that the loss of farms in the most agriculturally dependent counties is likely to be very extensive and concentrated among certain types of producers and farms.

The analysis presented below is based on a number of highly speculative assumptions, many of which may prove to be erroneous. As noted in the opening remarks in this chapter, we believe that tracing the likely impacts of the loss of such producers on agriculturally dependent areas (even if based on analysis dependent on speculative assumptions) is absolutely essential to the formulation of policies to address the implications and impacts of the farm crisis on rural America.

Implications for the Structure of Agriculture in Agriculturally Dependent Counties

One of the first questions that must be asked is how the structure of agriculture in agriculturally dependent rural counties may be changed by the rate of farm failure assumed in this chapter. The data in Table 7.6 provide one means of examining this question. The number of producers estimated to fail is 213,500, clearly too large to be drawn entirely from the agriculturally dependent counties in the United States. (This number is equivalent to 75 percent of *all* producers in agriculturally dependent counties in 1982.) Although such rates of farm failure will clearly not occur in agriculturally dependent counties, the number estimated to fail does illustrate that the rate of failure in the most agricultural areas of the United States is likely to be much larger than that for the United States as a whole. Whereas the 213,500 we estimate to fail is less than 10 percent of the 2.2 million farms in the U.S. in 1982 and less than 15 percent of the more than 1.5 million commercial farms (farms with at least $40,000 of sales) estimated by Johnson et al. (1986) to be in the United States in 1985, our estimates clearly show that the rate of failure is likely to be much higher in those areas of the nation where agriculture is concentrated. The proportion that will fail will be extensive, and such farms will disproportionately be in agriculturally dependent counties.

The data in Table 7.6 demonstrate that the producers likely to leave agriculture will be young, with farms involved primarily in grain production and producers farming middle-sized commercial farms. Thus, using data from Johnson et al. (1986) on farms by debt-to-asset ratios in different sales categories, by crops produced, and by producer age, and the assumptions noted above concerning failure rates in different debt-to-asset categories, we estimate that nearly 64 percent of the producers who will fail will be less than 45 years of age, 36 percent will be involved primarily in grain production, and nearly 28 percent in livestock production, and more than 50 percent will be in farms with sales of $40,000 to $249,999. It is evident, then, that the farmers likely to leave agriculture as a result of the crisis are those who operate the middle-sized farms in which America's family farms are concentrated.

The data in Table 7.6 clearly point to high rates of farm failure in agriculturally dependent counties. Although the rate in these counties is unlikely to be as high as estimated, it is evident that the proportions likely to fail in these counties will be high and that the farms that do fail will have distinct characteristics. These farms are likely to be operated by young producers who are operating middle-sized commerical-scale farms. If such estimates even approximate the actual events that occur over the coming years, the farm crisis will result in a further accentuation of the bipolar structure of agriculture and further increase the average age of America's farmers. The effects of the crisis on the structure of agriculture will clearly be extensive.

Demographic, Economic, Public Service, and Fiscal Implications for Rural America

In the discussion that follows, some of the potential implications of the loss of alternative numbers of producers and related (secondary) workers for the demographic, economic, public service, and fiscal characteristics of agriculturally dependent nonmetropolitan counties in the United States are examined. The reader is again reminded that the assumptions used in estimating these values are in many cases quite speculative. As a result, the values are best seen as indicators of the general rather than the actual magnitude of impacts that may occur as a result of the crisis.

Table 7.7 presents data on the demographic implications of the two alternative scenarios of population loss. As the data in this table indicate, the loss of the proportion of producers estimated here would result in the direct loss of more than 811,000 persons from agriculturally dependent rural areas. Under scenario I, an additional 165,000 would be lost from these areas and, under scenario II, an additional 496,000 would leave rural areas due to the loss of secondary workers whose jobs are dependent on agriculture. A total of between 976,000 and 1,307,000 persons are thus estimated to leave agriculturally dependent counties in the United States as a result of the farm crisis. These numbers represent 23.9 and 32.0 percent respectively of the 1980 population in these areas. A loss of between one-quarter and one-third of these counties' population bases will clearly have significant impacts on these areas.

The impacts on the populations of these areas, however, lie not only in the total numbers who will be affected but also in the manner in which the age structures of rural areas will be affected. The last panel of Table 7.7 shows the potential effects on the age structures of agri-

Table 7.7

Estimated Total Population Leaving Agriculturally Dependent Counties by Age and Its Potential Effect on the Age Structure of Agriculturally Dependent Counties

Total number of producers estimated to leave agriculturally dependent counties	213,500
Other producer-related dependents leaving areas	597,800
Scenario I	
Secondary workers leaving area	53,375
Secondary-worker-related dependents leaving areas	112,088
Scenario II	
Secondary workers leaving area	160,125
Secondary-worker-related dependents leaving areas	336,263
Total producer-related population leaving agriculturally dependent rural areas	811,300
Total secondary population leaving agriculturally dependent rural areas	
Under scenario I	165,463
Under scenario II	496,388
Total population (producer- and secondary-worker-related) leaving agriculturally dependent counties	
Under scenario I	976,763
Under scenario II	1,307,688

Total Population Change by Age in Agriculturally Dependent Rural Counties Due to the Loss of Alternative Number of Producers and Secondary Workers

		Scenario I		Scenario II	
Age Group	1980 Census Population in Ag-Dependent Counties by Age	Assumed Population Leaving by Age	Percent Population Change	Assumed Population Leaving by Age	Percent Population Change
< 18	1,230,971	373,625	−30.4	523,075	−42.5
18-34	998,743	215,729	−21.6	271,593	−27.2
35-44	414,438	165,616	−40.0	216,193	−52.2
45-54	403,456	122,947	−30.5	162,184	−40.2
55-64	422,219	74,479	−17.6	110,275	−26.1
65 +	610,038	24,367	−4.0	24,367	−4.0
Total	4,079,865[b]	976,763	−23.9	1,307,688	−32.1

[a]For a description of the assumption underlying these estimates, see the text and Table 7.5.
[b]The population shown here differs slightly from that in Table 7.1 due to the fact that age was not reported for 1,223 persons in these counties.

Table 7.8

**Estimated Economic and Business Related Losses Due to Alternative Number
of Producers and Secondary-Worker-Related Populations Leaving
Agriculturally Dependent Counties in the United States**

	Changes Under Alternative Scenarios	
Item	Scenario I	Scenario II
Bank deposits	$5,078,119,864	$6,798,578,374
Bank savings	797,350,764	1,067,491,864
S & L savings	822,365,652	1,100,981,758
Earnings ($) from:		
Agriculture, forestry, and fisheries	58,342,024	78,108,204
Mining	70,658,999	94,598,150
Construction	194,756,675	260,739,910
Manufacturing	499,008,426	668,071,645
Transportation and utility	253,176,840	338,952,730
Wholesale trade	211,049,073	282,552,146
Retail trade	390,968,726	523,428,276
Financial, insurance, real estate	131,462,465	176,001,728
Services	388,819,848	520,551,362
Government	719,151,158	962,798,367
Retail sales	2,907,411,722	3,892,437,947
Number of private sector business establishments	18,930	25,343
Total personal income	$7,630,527,256	$10,215,737,117

culturally dependent rural areas. The proportion of persons of less than 18 years of age leaving these areas would be between 30 and 42 percent, the proportion aged 18 through 34 between 21 and 27 percent, the proportion between 35 and 44 years of age, 40 to 52 percent, and for those 45 through 54 years of age, 30 to 40 percent would leave such counties. Cumulatively, between 28 and 38 percent of all persons under 45 years of age are projected to leave agriculturally dependent counties as a result of the farm crisis. For the remaining population the proportion of persons in the dependent ages of less than 18 and 65 years of age or older would constitute 46 percent of the total population, implying a nearly one-to-one ratio between dependent and other persons. The fact that more than 20 percent of the total population would be more than 65 years of age points to potential problems in the long-term maintenance of community organizations and to a heavy service burden for the remaining rural residents in the working ages.

Overall, then, the demographic impacts of the farm crisis will be significant for agriculturally dependent rural areas. The crisis will likely decrease the size of the population by 25 percent or more, result in the loss of a significant number of persons in the productive working ages, and leave a remaining population in these areas that is increasingly dependent and elderly. Such patterns suggest that the demographic implications for agriculturally dependent rural areas are indeed extensive.

Some of the potential implications of the loss of a substantial number of producers for the economic and business sectors of agriculturally dependent nonmetropolitan counties, are evident in Table 7.8. In examining the results in this table, it must be recognized that the impacts stated may substantially underestimate or overestimate the actual impacts. The fact that we believe that our estimate of the number of producers likely to fail is conservative suggests that the impacts shown are likely to substantially underestimate the actual impacts. On the other hand, much of the production-related spending presently being done by producers who will eventually leave these counties, will still be done by other producers who will take over the operation of the farms. The new operators of these farms will be required to spend similar amounts per acre to produce crops from the same acreage. Although we have attempted to take this into account by assuming relatively small secondary impacts, it is essential to note that some overestimation of economic effects could occur.

Although the use of per capita rates for economic variables results in the impacts for each indicator being proportional to the total population loss in agriculturally dependent nonmetropolitan counties (i.e., 23.9 percent under scenario I and 32.0 percent under scenario II), it is nevertheless important to examine the magnitude of such losses. As an examination of the data in Table 7.8 suggests, the results for the economic sectors of agriculturally dependent counties would be substantial. Bank deposits would decline by between $5 and $6 billion, bank savings by between $800 million and $1.1 billion, and savings and loan savings by between $800 million and $1.1 billion. Although persons facing severe financial problems are unlikely to have savings or other funds of this magnitude, as noted in chapter 5, they are likely to have financial impacts of similar magnitude as a result of unpaid debts that will remain after they leave rural areas.

Earnings in the various sectors of the economy would also decline. Trade and service centers would clearly be impacted because of the loss of producers, secondary workers, and the dependents of producers and secondary workers. Wholesale trade could decline by nearly $300 million, retail trade by more than $500 million, and services by more than $500 million. In addition, although it is difficult to estimate the exact loss in government earnings because many of the changes in government earnings are not affected by population change, it is apparent that the potential loss in government earnings could be extensive.

Finally, it is evident that the overall losses in retail sales, business establishments, and income would be substantial.Retail business volume could decline by from $3 to $4 billion, between 18,000 and 25,000 business establishments could be lost, and between $7 and $10 billion in income could be lost from agriculturally dependent areas as a result of the process of farm failure and the related decline in the number of producers and secondary workers residing in these areas.

The economic impacts estimated as potentially occurring as a result of farm failure would clearly be significant for rural communities and their residents. A dramatic loss in capital to fund development projects, substantially reduced earnings to be spent in the local area, and the loss of a large proportion of retail trade expenditures and businesses may result in a sharp decline in many rural trade centers. For many such centers, it will no longer be possible to provide even the minimum range of services required by their population bases, and as a result, their remaining residents will be forced to travel farther to obtain needed goods and services. Such levels of decline clearly suggest that the loss of business involved in the estimated loss of producers could lead to the decline of many small communities. In fact, given that the average rural community in 1980 consisted of roughly 800 persons (U.S. Department of Commerce 1983), the population loss estimated above, coupled with the decline in business volume noted here suggests that the current crisis could lead to the equivalence of the death of more than 600 rural trade centers during the coming years (assuming each center consists of 800 people and that 496,388 secondary-worker-related persons will leave rural areas).

Table 7.9 provides data on the potential impacts of the estimated population loss on public services and on the fiscal characteristics of agriculturally dependent nonmetropolitan counties. Although the overall proportional losses are again largely equivalent to the proportional loss in population, it is useful to examine the magnitude of the impacts. Again, the reader should be aware that some underestimation or overestimation of impacts may occur given the assumptions used in the analysis.

In particular, because much of the local tax revenue generated in rural areas is due to property taxes, which will continue to be collected after the producers leave, the loss in tax revenues is likely to vary with the changing value of rural property. Because the amount of decline in rural property values is so variable from one area to another and because it is nearly impossible to project losses due to the decline in property values for the nation as a whole (Stinson et al. 1986), estimates of local revenue loss have been adjusted to eliminate the effects of potential loss due to changes in property values. This was done by reducing total values by 30 percent because the average proportion of local revenues coming from property taxes in agriculturally dependent rural areas was 30 percent in 1982 (as computed by the authors from the 1982 Census of Governments). Despite this adjustment, it is still unlikely that the amount of revenue will decline linearly with the population decline in the manner indicated in our estimates, and thus some overestimation of revenue loss may be evident in our estimates. On the other hand, the decline in rural land values will substantially affect rural tax revenues (Stinson et al. 1986). As a result, the actual tax revenue loss occurring in rural areas may be substantially larger than that estimated here.

An examination of the data in Table 7.9 shows substantial public service and fiscal impacts resulting from the farm crisis in agriculturally dependent rural areas. Service demand could decline sufficiently to eliminate the need for from 1,300 to 1,800 police officers, and between 37,000 and 50,000 full-time, local government employees could lose their jobs due to the decline in demand for government services. In addition, the number of physicians needed could decline by nearly 700, the number of hospital beds by more than 5,200, and the number of social workers by more than 650. Among the most notable losses could be the loss of between 200,000 and 275,000 school children. A decline of this magnitude will clearly result in the closing of a large number of small rural schools. Finally, although it may initially appear that the reduction of expenditures resulting from reduced population size will more than offset the loss in tax revenues, it is clear that service costs

Table 7.9

**Estimates of Public Service Demand and Fiscal Losses Due to Alternative
Numbers of Producers and Secondary-Worker-Related Populations Leaving
Agriculturally Dependent Counties in the United States**

Service/ Fiscal Item	Standard Assumed Values*	Scenario I	Scenario II
Police officers	1.37	-1,338	-1,792
Police vehicles	.66	-645	-863
Public school students	.21	-205,120	-274,614
Full-time government employees	38.35	-37,459	-50,150
Physicians	53.12	-519	-695
Hospital beds	4	-3,907	-5,231
Gallons of water	150	-146,514,375	-196,153,200
Miles of highway	.1	-97,676	-130,769
Number of social workers	.5	-488	-654
Number of firefighters	2	-534	-747
Local government general revenue (excludes property taxes)	$780.70	-$762,558,874	-$1,020,912,022
Government expenditures	$1069.59	-$1,044,735,402	-$1,398,690,008
Debt outstanding	$673.45	-$657,800,706	-$880,662,484

*Standards used in estimating service and fiscal demands were drived from
the U.S. Bureau of the Census, CO-STAT 2 tape (U.S. Department of
Commerce, 1986a). Standards for police officers, students, government
employees, physicians, hospitals, government revenues, expenditures and
debt are from Murdock et al. (1987b). Standards used were as follows:
police officers per 1,000 population = 1.37; police vehicles per 1,000 =
.66; public school students per capita = .21; full time government
employees per 1,000 population = 38.35; physicians per 100,000 population
= 53.12; hospital beds per 1,000 persons = 4; gallons of water used per
person per day = 150; miles of highway per person = .1; number of social
workers per 1,000 people = .5; number of fire fighters per 10,000 = 2;
local government general revenue per capita (excludes property taxes) =
$780.70; government expenditures per capita = $1,069.59; debt outstanding
per capita = $673.45.

do not decline in a linear manner with population (Doeksen and Peterson 1986). It will be impossible for some school systems and for many other types of services to simply reduce service levels as populations decline because many services have high fixed costs (e.g., costs for school buildings and teachers to teach courses required by state curriculum standards) that cannot be reduced. In addition, it is evident that the persons leaving rural areas will leave a substantial level of debt in local areas. If these levels of debt are added to the estimated decline in revenues, the loss of the number of producers and other residents projected here will cause fiscal problems for rural areas.

Analysis of these data clearly suggests that rural service and fiscal bases in agriculturally dependent rural areas will experience difficulties due to the estimated loss of between 970,000 and 1,300,000 producers and related workers and dependents. These impacts will include not only the loss of the level of service demand necessary to support between one-quarter and one-third of existing service capacity but also the loss of the revenue base necessary to support existing services and an increasing level of per capita debt for the remaining residents of such areas. These potential impacts point to the need for substantial adjustments in the service and fiscal bases of such areas in the coming years.

The Social Implications of the Crisis

Finally, it is critical to examine some of the social implications of the loss of a large proportion of producers and other rural residents from agriculturally dependent nonmetropolitan counties. Although the tracing of such impacts can be, at best, inexact (see Murdock et al. 1986), it is nevertheless essential to discuss the potential social changes likely to occur as a result of the current farm crisis.

In many cases, social impacts are the result of changes in other basic socioeconomic factors. In regard to the current farm financial crisis, we believe the social impacts of the crisis are likely to result, in large part, from the loss of a substantial proportion of young adults, from the loss of economic capital and resources, from the fact that the loss appears to disproportionately involve well educated and civicly and socially involved residents of rural areas, and from the fact that individual producers and other residents of rural areas are experiencing social stress and related problems that will not be easily resolved.

The loss of the proportion of young adults forecast in this chapter will result in both demographic and social disjunctures. Demographically, as noted above, such losses may lead to relatively old age distributions among rural populations and to further accentuate existing high rates of old-age dependency in these populations. Socially, the loss of young adults is problematic because it is primarily from such persons that new ideas, new approaches, and new resources for further community development are introduced into rural areas (Rogers 1983). It is also these persons that are most likely to support social changes such as moves towards greater equality (e.g., between the sexes or racial/ethnic groups) and the introduction of new agricultural and other forms of technology (Murdock et al. 1986a). The fact that young adults will disproportionately be lost from agriculturally dependent rural areas means that for both agriculture and other sectors of rural society the chances of social stagnation increase.

The fact that the loss of producers is likely to be accompanied by an accompanying and substantial loss in economic capital and resources is also of social significance. The socioeconomic conditions of rural areas, as noted in chapter 3, are already disadvantaged relative to other areas in the United States. They can ill-afford the loss of additional resources. The loss of economic resources from these areas will severely limit the range of alternative forms of economic development that can be pursued, because of the costs associated with the development of multiple alternatives, and will limit the potential for social mobility within the social structure of agriculturally dependent areas. The loss of resources from such areas will thus likely limit the ability of these areas to pursue forms of economic development that may serve as alternatives to their declining agricultural base.

The fact that those being lost are, as shown in chapter 6, active in the leadership and membership of the major social organizations of rural areas is also important. These organizations form the backbone of social life in rural communities (Christenson and Robinson 1980) and serve to provide the social psychological and social support for rural residents during times of change and tragedy. They also provide substantial financial support for persons in need of assistance and for general community growth and development. The loss of persons actively involved in such organizations will not only result in a decline in a vital part of the social organizational and social interactional fabric of rural America but will also accentuate the magnitude of loss in rural areas because the persons leaving the area represent a loss of not only individual social interactants but also major participants in the organizational life of the community. Finally, as community organizations are weakened by the loss of key members and leaders, the capability of such organizations to serve as bases of financial and social support for persons under stress diminishes substantially. A decline in the strength of the organizational base of rural areas is a likely major social impact of the current financial crisis.

Finally, it is evident that a major social psychological impact of the current farm financial crisis is that many rural residents are experiencing a distancing from family members as a result of their forced movement to seek alternative forms of employment or through crisis-induced divorce and other marital and family discord. In like manner, many persons are experiencing major career failures that will create formidable social and psychological problems for such persons to reconcile in the years to come. Although it is impossible to adequately quantify the costs of social stress, it is apparent that these costs are extensive and that such stress has permanently affected the lives of many of the producers and other rural residents being affected by the crisis. Such persons are likely to be less effective members of the social order and may, in fact, require substantial public investment in the resolution of their problems (e.g., funds for unemployment compensation, for psychological counseling). The costs of social stress then are apparent and real in both human and economic terms (Heffernan and Heffernan 1985a, 1985b).

In sum, the long-term social implications of the current financial crisis in agriculture for agriculturally dependent nonmetropolitan counties include a decline in the number of purveyors of innovation for new methods and technologies, a likely tendency for the social hierarchy to stagnate, a reduction in the potential for the area to pursue new and diversifying forms of alternative types of economic development, a decline in the size and vitality of community organizations, and a legacy of a relatively large number of persons with permanent social and psychological problems. The social impacts may be among the most costly impacts of the farm financial crisis in rural America.

Summary and Conclusions

In this chapter, the potential long-term implications of the loss of a substantial proportion of agricultural producers on agriculturally dependent nonmetropolitan counties have been examined. Implications were examined for the structure of agriculture; for economic, demographic, public service, and fiscal factors; and for the social characteristics of such rural areas. Although the impacts estimated are highly speculative, if the actual impacts of the crisis even approximate those projected, these areas will be substantially and permanently changed by the current crisis. They may lose more than 213,000 farms, and more than 60 percent of these will be middle-sized farms with sales between $40,000 and $250,000. They may lose more than one million, or nearly one-third, of their residents, a substantial part of their economic and trade base, and more than 200,000 school children and will experience a declining fiscal base to support needed public services. They may also lose the vital base of future leadership essential to guide such areas in alternative paths of development. The loss in population and the related economic and service support would be equivalent to the loss of more than 600 rural service centers in the United States. The implications of the current farm crisis for the long-term future of rural areas are substantial.

Although the effects within each of the areas of impact (e.g., agriculture, population, business) noted above are extensive, the implications for agriculturally dependent counties are even more sobering if they are examined as they are likely to collectively impact the common socioeconomic base of agriculturally dependent rural areas in the United States. Examined as a whole, the changes described in this chapter present a picture of agriculturally dependent counties as areas with a declining number of producers, particularly in mid-size farm categories, and a small and dwindling base of young producers. It further points to areas with decreasing populations, populations that are generally old and getting older, and populations with a very small number of young adults. These are areas with declining business sectors, a rapidly declining base of economic opportunity for their residents, a decreasing range of services, and decreasing resources to support such services. Finally, the picture is one of communities with a declining number of organizations, declining membership within organizations, and organizations in which critical leadership shortages may occur with increasing frequency. Taken together, it appears that many agriculturally dependent areas may experience such pervasive patterns of decline that the human resource base essential to redevelop these areas could be permanently depleted. With rural America already bearing a disproportionate share of the nation's poverty, unemployment, and similar disadvantages (Deavers et al. 1986), it appears that among the major implications of the current farm crisis may be a further accentuation of the problems of poverty and economic underdevelopment that already plague rural areas.

In sum, we believe that the long-term implications of the current farm financial crisis are a cause for considerable concern, when placed within the context of a rural America that has already had its farms, its businesses, and its people depleted by decades of decline. As we note in the following chapter, it is critical that policies be formulated to address what may otherwise be a further accentuation of a long-term pattern of economic and social decline in rural America.

Policy Alternatives
and Research Agenda

Steve H. Murdock and F. Larry Leistritz

In this volume we examined the context of the current farm financial crisis, described the impacts experienced by producers and by residents in rural communities, and examined possible long-term implications of the crisis for agriculturally dependent rural areas in the United States. The analysis has revealed that the crisis has occurred in a context that already limited the alternatives for rural residents; that the impacts of the crisis have been extensive for producers, other rural residents, and rural communities; and that the long-term implications of the crisis may be substantial, resulting not only in the loss of a large proportion of the existing population base but also depleting the human and economic resource base of such areas to levels that may make alternative forms of economic development difficult.

Among the specific impacts of the crisis are increased rates of farm failure, increased farm debt, and reduced assets for those who remain in farming. For the agricultural base of rural areas, the crisis is resulting in an accentuation of the rate of loss of middle-sized farms and in the loss of some of the best-educated and innovative producers. For rural communities, this crisis is leading to the failure of many small businesses and to economic hardship for both business proprietors and wage earners in rural communities. Both producers and other residents of rural areas are experiencing significant personal problems because of the crisis. Rural areas may also lose a substantial proportion of their leadership and the membership base for the community organizations essential to the functioning of these communities. In sum, the impacts are substantial and pervasive.

Over the long-term, these impacts may lead to the loss of more than one million persons from agriculturally dependent rural counties, to the deterioration of the base of services and financial and other resources in agriculturally dependent rural areas, and to the further deterioration of socioeconomic opportunities in rural areas. That the base of the potential population likely to be affected is small relative to the nation or to even the population in the rest of nonmetropolitan America (in fact, it is less than that in many of the largest U.S. cities) and that U.S. society as a whole may experience few impacts as a result of the changes occurring in such counties must be recognized. The impacts are sufficiently apparent and extensive, however, to require careful consideration by those who are concerned about rural America and who believe that an essential part of America's heritage may be tied to the future of rural America. For such persons, the impacts and the implications of the crisis require the development of policy alternatives to address the needs of those who are being impacted by the crisis and to reverse long-term trends that threaten to limit the base of opportunities for those who live in rural areas in the United States. The major topics of this final chapter are the examination of policy alternatives for addressing both the short-term and long-term implications of the crisis and a discussion of research that must be completed to inform policy making.

Policy Alternatives

The policy discussion presented addresses the problems of producers resulting from the crisis and the problems both directly and indirectly related to the crisis for agriculturally dependent rural counties. No attempt is made to describe farm policy alternatives as they relate to commodity programs, land set-aside programs, price-support programs, etc., nor is an attempt made to examine U.S. monetary and trade policy although changes in such policy areas have affected and will clearly continue to affect the socioeconomic conditions of agriculturally dependent rural areas. Descriptions of such policy alternatives are available in numerous sources (Tweeten 1970; Knutson et al. 1983; Paarlberg 1980) and are not examined here. We also do not attempt to discuss long-term economic development and diversification strategies for rural areas (see Tweeten and Brinkman 1976; Lonsdale and Seyler 1979; Drabenstott et al. 1987; Henry et al. 1986). Finally, we do not address the issues surrounding the policy formation and implementation processes. These processes are complex and involve the interplay of numerous social, political, and interest group concerns. Clearly, these processes will determine the feasiblity of enactment of any proposed policy. However, these dimensions are adequately discussed elsewhere (Long et al. 1987) and are beyond the scope of the present effort. The focus in this chapter is only on policies that might be pursued to immediately address the direct and indirect results of the current farm financial crisis, particularly in agriculturally dependent counties.

Of the many aspects of the crisis toward which alternative policies may be addressed, three are particularly relevant. These are programs aimed at directly assisting producers, programs aimed at assisting business operators and other persons in rural communities, and programs aimed at strengthening the service, resource, and organizational bases of rural communities. Potential policies to address each of these areas are delineated below.

The No-Action Alternative

Any discussion of policy alternatives would be remiss if we did not first discuss the no-action option of allowing market and other economic forces to simply take their course. This alternative, in fact, has been the basic policy followed in the past and appears to be supported by the history of social change in rural America. It can be argued, for example, that the loss of large numbers of producers is not a new phenomena and that the subsequent effects on rural communities have historically been absorbed without the need for basic changes in policy. In addition, it can be argued that with an increase in farm income in 1986 and in 1987, the events necessary to resolve the crisis are already occurring. These "it is too late" or "it is now unnecessary" arguments fail to take into account either the breadth or depth of the impacts of the crisis or their ramifications for rural nonfarm residents. The impacts of the crisis are still occurring, and in fact, we argue the impacts on rural communities are becoming increasingly apparent at the time of this writing and will continue for years.

Whether policy actions are necessary, however, depends upon what is likely to happen in the absence of any actions and on the extent to which one values those factors likely to be negatively impacted by the lack of a concerted set of actions. Our bias in regard to the preservation of rural areas has already been acknowledged. In addition, it is apparent that the present situation is different than those which have prevailed in the past. Our examination of the context of the crisis suggests that agriculturally dependent rural areas in the United States will not retain their ability to provide quality living environments and maintain the potential to pursue other forms of economic development if policy changes are not implemented. Although such a view may be seen as an alarmist's perspective, we see little to suggest that rural areas presently have the *excess capacity* necessary to allow them to absorb the impacts of the present crisis without incurring costs that may permanently alter their long-term potential for economic growth.

In sum, we argue that the no-action alternative is not a feasible alternative if agriculturally dependent rural areas are to retain their viability and their potential to respond to opportunities for renewed economic development. The policy options discussed below are ones that should assist rural areas in retaining that viability and the potential for future development.

Policies to Assist Producers

A wide variety of means might be used to assist producers through the current crisis. Among these options (Leistritz et al. 1985), which are discussed in greater detail below, are:

1. direct financial assistance programs aimed at reducing producers' overall levels of debt;
2. programs aimed at assisting producers in renegotiating loans and in altering payment schedules;
3. the provision of assistance to producers in improving their management and decision-making skills;
4. assistance to producers in retraining to obtain nonfarm employment and assistance in relocating to alternative job sites;
5. the provision of mental health and other human services.

Direct Financial Assistance. One of the apparent solutions to the farm crisis is that of directly assisting producers in making interest and principal payments on farm debts. This option is the most direct route to alleviating the major cause of farm failure. The major difficulty with this option is the potential costs involved. Although no national-level analysis has been conducted, the authors have examined the costs of providing funds sufficient to allow producers in debt to pay operating expenses, to make principal and interest payments on long-term and intermediate-term debt (assuming long-term debt is amortized over 20 years and intermediate-term debt over 5 years), and to allow producers' families to live at the poverty level (as defined in 1985) *for a single year*, given that farm income was at the 1985 level and farm expenses and debts were distributed according to the results for respondents in the 1985 surveys of producers in the two states. For North Dakota, the estimated annual costs would be $375 million (the entire general revenue budget of the State of North Dakota in 1986 was $563 million) (Leistritz et al. 1985a) and $674 million in Texas. North Dakota and Texas, like many states experiencing the crisis, are having statewide budget problems, and such expenditures are clearly unlikely to be approved. The costs at the national level would be staggering. Given such high costs, the option of simply providing direct financial assistance to producers to assist them out of their current debt problems does not appear likely to be supported by either the general public or state or Federal decision makers.

Numerous states are making attempts to implement related programs, however. A recent summary by Popovich (1987) showed that 21 states had initiated programs by the end of 1986 aimed at alleviating the financial stress of producers. These states had authorized $1.6 billion for a variety of programs aimed at assisting producers. The programs implemented by states include linked-deposit programs (e.g., Illinois, Indiana, Kansas, Missouri, Montana, and Ohio) in which the state provides low interest money to lenders in exchange for them providing low interest loans to producers; interest rate buy-down programs (e.g., Iowa, Wisconsin, and Minnesota) in which the state pays for the interest on producer loans above a prescribed level with no expectation of producer repayment; interest rate deferral programs (e.g., Illinois and North Dakota) in which the state pays interest above a prescribed level for several years with the producer being required to repay these costs at some later date; interest deferral/forgiveness programs (e.g., North Dakota and Minnesota) in which the state pays the interest and the producer repays these costs, only if the value of the items covered by the loan show specific patterns of change; farm grant programs in which the producer is provided with a direct grant (e.g., Alaska) to pay debts; and tax credit programs (e.g., Kansas and Idaho) in which the state provides tax credits to lenders on interest earned on producer loans. Although it is too early to discern what effects such programs may have, it appears that although such programs may assist some producers in surviving short-term financial crises, it is unlikely that they can alleviate the debt problems in agriculture given that the $1.6 billion in funds authorized are so small relative to the estimated producer debt of more than $157 billion as of January 1, 1987 (Johnson et al. 1987). Nevertheless these programs do indicate that decisionmakers are seeking means to address the crisis.

Renegotiating Loans. Perhaps more feasible is the possibility of assisting producers in renegotiating current interest rates and debt payment schedules. The State of Minnesota, for example, has used a system of state-supported mediators to meet with producers and lenders to negotiate loan payment extensions. Such situations are often desirable not only for the producer but for the lender as well. Many lenders with loan portfolios, which include numerous high-debt producers, have very little to gain from the foreclosure of a farm that may be difficult to resell and for which any resale will be at a substantially lower price and be financed at a lower interest rate. If the existing owner is able to pay the debt, even if the repayment period is longer than originally negotiated, the lender may still obtain a higher return than that obtained through foreclosure. However, as shown in chapter 6, the majority of producers do not favor the option of direct financial assistance programs, and many do not favor interest buydowns or even loan renegotiations. Such an option would have to be pursued with full recognition of the likely opposition and the need to ensure equity to borrowers who are keeping current on their debt payments. It represents, however, an important option for alleviating the financial burdens of some producers.

Improving Management and Decision-making Skills. Other producers, particularly those with intermediate levels of debt, may be assisted by helping them to improve their management and decision-making skills. Although such groups as the Cooperative and Agricultural Extension Services in many states (e.g., North Dakota, Iowa, Nebraska, and Missouri; see Mazie and Bluestone, 1987) have initiated such programs with considerable success and although general programs of this nature have formed a major component of Extension education programs for years, the initiation of broader nationally oriented programs could be beneficial. The fact that the producers with the highest levels of debts relative to assets are relatively well educated should increase their ability to utilize such programs.

Retraining and Relocation Assistance. For many producers, the reality is that they will be forced to leave farming and pursue new nonfarm careers. Because many of those who will fail are at mid-career stages with families that are at the most dependent ages, such transitions may be particularly difficult. These producers need retraining and assistance in meeting family expenses during the period of retraining and may need financial and other assistance for relocating to new areas. Again, the fact that the producers likely to fail in farming are relatively well-educated promises to make them excellent candidates for retraining. Consideration should be given to the provision of low-interest and no-interest loans to producers to assist them in retraining. As noted in chapter 6, this is a popular concept, even among those producers with low levels of debt who otherwise were not in favor of direct aid to producers, and thus should be a generally popular policy option. Un-fortunately, such programs have been pursued in only a few instances (e.g., Section 1440 Displaced and Distressed Farm Family Programs implemented in selected states). Nationwide programs of this nature are being pursued in Canada, however, and are considered to be relatively good investments compared to the costs that may be incurred if producers are forced to access various types of public aid and assistance (Canadian Department of Employment and Immigration, 1986). Consideration should be given to such programs for agricultural producers and for other technologically and otherwise displaced U.S. workers.

Mental Health and Human Services. One of the apparent needs of producers, and of other rural residents as well, is the need for additional assistance for personal, family, and marital counseling and for other human services that have historically been inadequate in rural areas (Ulman and Olson 1984).The need for at least short-term enhancements of such services is evident in the results of our analyses and those of others (Bultena et al. 1986; Heffernan and Heffernan 1985a), which show very high incidences of socioemotional problems among producers and other rural residents. In addition, available evidence suggests that producers in many areas have been unwilling to solicit assistance from, or have found traditional sources of assistance such as rural churches of limited utility in meeting their needs (Heffernan and Heffernan 1985a). It is also evident that the close, intimate, and public nature of the interaction patterns in many rural areas makes concerns related to the labelling sometimes associated with seeking such services a barrier to many producers needing assistance. Finally, such assistance is often seen as somehow less important than financial and other economic forms of assistance so that it is often omitted from otherwise comprehensive aid packages. Given the high frequency of socioemotional problems among producers and others being impacted by the crisis, and the potential long-term impacts of such problems, such services should not be seen as frivolous and less important than economic assistance. In fact, the successful resolution of the feelings of inadequacy and failure that often accompany career failure may be as important as economic packages of assistance to the successful adjustment of former producers and to the motivation of rural residents to seek new development options for their communities. The provision of such services is essential.

Summary. A number of options have been suggested to directly assist producers. Although some options, such as direct financial assistance or buy-down programs do not appear feasible, serious consideration should be given to the expansion of programs aimed at assisting producers and lenders in negotiating mutually beneficial financial arrangements, in increasing the range of services in business and management training available to producers, in assisting producers to retrain and relocate, and in obtaining additional counseling and other forms of assistance in resolving socioemotional problems. Although such programs may seem extensive and expensive, it is unclear how inexpensive the option of simply allowing producers to adjust to the crisis without assistance may be, given their likely need to access a variety of types of public funds even if no specialized assistance is provided. As we shall argue below, it is essential that research be conducted to determine the relative costs and benefits of alternative options, but until such research can be completed, it seems prudent to pursue options that *may* be cost effective and *may* allow producers and other displaced workers to again become productive members of the workforce as soon as possible.

Policies to Assist Other Rural Residents

As the discussion in this volume has indicated, rural business operators and other rural nonfarm residents are being impacted by the crisis as business activity has been reduced and related employment needs have declined in rural communities. For such persons, many of the needs are similar to those of producers; that is, business operators need assistance in managing the debt related to their operations and assistance in improving their management skills. Both business operators who are forced to quit business and employees in rural areas need assistance in retraining and relocation to sites where job opportunities exist, and both are experiencing levels of socioemotional problems that show a need for increased counseling and related human services.

In the process of policy formation, then, it is essential that the same alternatives as noted above for producers, including potential loan and payment renegotiations, financial and business management training, vocational retraining, residential relocation, socioemotional counseling, and human services be formulated to include such alternatives for nonproducer groups in rural areas. Because the impacts on nonproducer groups are less directly attributable to the farm crisis than those for producers, and because it is often difficult to separate impacts due to the farm crisis from those due to other causes (e.g., declines in business activity due to reductions in rural manufacturing or in gas and oil exploration activity), policy formulation often ignores the needs of the nonfarm component of rural areas. The interdependence of farm and nonfarm activities in rural areas is sufficiently complex, however, to make attempts to separate farm and nonfarm impacts extremely difficult and unnecessary if the goal is one of alleviating the impacts of the crisis on rural America (Korsching and Gildner 1986). For rural America, policy options aimed at alleviating the impacts of the crisis must include the total rural community of producers as well as other rural residents.

Policies to Assist Communities

The farm crisis also has impacts on rural communities in addition to those that occur for individuals in the communities. For example, the impacts on community service and fiscal bases may affect the ability of the community to compete for economic development projects and to provide an adequate range of services at reasonable costs for residents who remain in rural communities. In order for many rural communities to survive as viable rural trade and service centers, we believe it is essential to develop policies to:

1. maintain and enhance rural community infrastructures;
2. increase part-time and fulltime job opportunities in rural communities;
3. promote cooperative programs for economic development and service provision among rural communities;
4. develop management expertise and to support community organizations and institutions.

In general, of course, programs to address such dimensions are already being conducted by the Cooperative Extension Service, the Farmer's Home Administration, the Economic Development Administration, the Rural Electrification Administration, and similar agencies (Long et al. 1987), but enhancement of the funding for most of these programs is essential. In addition, however, the present situation appears to call for efforts beyond the normal range of activities conducted by these entities. For example, in many states the major source of human and mental health services is the state. Since many of the states where the crisis is having its most substantial effects are ones that are facing substantial budgetary shortfalls, it is unlikely that increases in funding will be forthcoming for such services from local or state sources. Supplementary funding from federal agencies must be granted, if service needs are to be met in many rural areas.

In addition, several of the need areas are ones that have largely been outside the areas of responsibility of the agencies noted above. Although the Cooperative Extension Service and other agencies have numerous programs aimed at improving rural leadership and strengthening rural families, these programs tend to be oriented to teaching general skills to individual leaders and families rather than to assisting these institutions in general. It may be necessary to focus such programs and to develop yet additional programs aimed at assisting rural institutions in specific areas. For example, municipal service management and fiscal management of charitable institutions (which play such a large role in the provision of services in rural areas) might be more useful areas for training in many rural communities than general courses in leadership. Although such focused programs have historically not been implemented in rural settings because the small numbers of persons managing such services in rural areas are insufficient to merit specialized courses, it is, in part, the lack of such specialized training that has prevented many rural areas from developing the necessary service delivery systems and placed them at a disadvantage relative to more urban areas in terms of service management. It may well be that specialized training in a few vital service areas would be more beneficial in strengthening the management of services in rural areas than more general programs.

In like manner, policies must be developed to assist rural institutions in managing transitions to changing socioeconomic and demographic conditions. In this regard, it appears that the promotion of proposals to combine service entities within common facilities, to merge complimentary services, and to promote multi-agency service integration (i.e., to ensure that duplication of costs for basic facilities are not being incurred because of historical jurisdictional differences or other reasons) might be pursued (Rogers and Glick 1973) for those services for which such consolidation can be shown to be cost effective (Doeksen and Peterson 1987).

Finally, it is essential that joint community efforts aimed at general regional economic development be promoted. The competition between rural communities for various types of economic developments has generally been counterproductive (Lonsdale and Seyler 1979). Given the limited resources of such areas, rural regions must work cooperatively to promote the development of their areas, and policies must be developed that promote such cooperative programs. In this regard, regional economic development foundations might be developed to receive and administer funds for region-wide programs for economic development (Stohr 1986). In such programs, the need to consider developments that produce both full-time and part-time employment should be considered because of the important role of part-time employment in retaining producers in agriculture (Albrecht and Murdock 1984). It appears, for example, that one of the reasons for the higher historical rates of farm failure in the Midwest and Great Plains has been that such areas offer fewer opportunities for off-farm work to supplement farm income and defray operating expenses (Albrecht and Murdock 1985a).

In sum, policies must be developed to address the immediate impacts of the farm crisis for producers and for other rural residents and for the rural communities that provide trade facilities and services to rural farm and nonfarm residents. Such policies must take into account that the crisis is a rural crisis, not just a farm crisis, and carefully include rural community residents as well as producers among those for whom new and expanded programs must be developed. In addition, the policies developed must be ones that not only recognize that the crisis encompasses both producers and other rural residents alike but also recognize that the crisis is a social, psychological, and perceptual crisis as well as an economic crisis. Steps to address these dimensions is of vital importance to the health of rural communities. Finally, any programs developed must attempt to integrate intracommunity services, promote intercommunity cooperation for general community development, and promote specialized training to improve community management and administration. The policies developed, then, must not only be different in magnitude but different in kind than those which have often been developed for rural areas in the past (Long et al. 1987).

Research Agenda

One of the keys to effective policy formation is an adequate information base on the effects of alternative courses of action, but as often noted in this volume, the farm crisis has raised a number of questions that the existing base of information and data is ill-equipped to address. In this section, we outline some of the types of research and some of the research questions that we believe must be addressed, if an adequate base of knowledge is to be obtained to guide policy formation and decision making. In addition, we examine several major conceptual and disciplinary questions that we believe are important for the advancement of our base of knowledge concerning the short-term and long-term implications of the farm crisis and other forms of economic decline. The types of research described and the questions noted are not intended to be inclusive but rather to emphasize only some of the most important types of research essential for policy formation and the development of the base of knowledge in the social sciences. The area is a fertile one for research and numerous additional issues of relevance will be delineated by other researchers. Our discussion here is thus intended to be exemplary rather than exhaustive.

Although it is an often noted research need, it is clear that the need for longitudinal studies is especially important in research on the impacts of the farm crisis. It is essential to follow samples of producers and other rural residents over time in order to answer some of the most basic questions related to the crisis. Without such research, it is impossible to definitively determine the answers to such questions as: How many producers at which debt-to-asset ratios will actually fail in farming? How many producers who fail have been able to obtain employment in the areas where they operated their farms and how many have been forced to relocate to find employment? How many business operators have been forced to cease operation of their businesses due to the crisis, how many of them and their employees will find employment in local areas, and how many will be forced to relocate to find employment? How long does it take for the impacts of a crisis to affect an area, and what is the duration of such effects? What are the lasting effects of a major financial crisis on a rural community? How effective have aid programs been at retaining producers in agriculture?

Only longitudinal analyses can address such questions, but such analyses must be carefully designed. Longitudinal studies of the crisis should include both producers and other residents of rural areas. The samples of producers and residents included should be large enough to allow for generalizations regarding such effects in areas with different degrees of dependence on agriculture and different levels of off-farm employment opportunities. The sample should also allow for generalizations regarding impacts in agricultural areas that differ in the major commodities produced by their farm firms and that differ in the patterns of farm ownership; in farm size; in the form of agriculture (e.g., irrigated and nonirrigated) being practiced; in the age, educational, and technology use characteristics of producers; and in other regards related to the structure of agriculture and the characteristics of producers. Although such studies are likely to be costly and difficult to complete, they are absolutely essential.

It is also essential that analyses be completed to investigate the actual levels of interdependence between agriculture and the nonfarm economy of rural areas. In too many analyses, such as that reported in chapter 7, it is necessary to simply assume levels of local expenditures, rates of leakage, etc. that result from agricultural expenditures, but such assumptions are based on an insufficient data base. Comprehensive information to establish the magnitude and complexity of such linkages must be obtained. Such analysis should include not only the standard trade-area analysis and analysis of producers' purchasing patterns for different items but also linked producer-business analysis in which comprehensive records of producers' expenditures by type and location are linked with business operators' records of customer sales and their own purchasing patterns in regard to other businesses. Such studies are likely to be extremely expensive and will require extensive time periods to complete; but until such studies are completed in a variety of settings, the extent and characteristics of farm-nonfarm linkages will remain largely speculative.

Yet another area of research where additional emphasis is necessary is in regard to producer- and business-operator adjustments. If programs are to be designed to assist farm and business operators in adjusting to financial crises, it is essential to have better information on which adaptive strategies are most likely to be effective. Longitudinal analyses are again essential, but care must be taken to ensure that a sufficiently large sample of operators is included so that the numerous factors that affect survivability can be adequately analyzed. In addition, it may be necessary to use approaches in which operators are asked to maintain records of a specified type over time. Although the use of such a procedure could bias the results of an analysis by serving to alter operator behavior, the use of other methods often leads to an over-reliance on data from those with the best financial records—which is also likely to bias the results obtained.

We also know very little about the processes of adjustment to the noneconomic dimensions of the crisis. Although information is available (as noted in chapter 6) that shows that producers and other rural residents have experienced such impacts as increased levels of stress, depression, family conflict, and similar problems, and some initial work suggests that traditional sources of assistance have not been widely used by those experiencing these impacts (Heffernan and Heffernan, 1985a), the actions taken by producers to manage such problems have not been adequately documented. What services and service providers did producers and other rural residents utilize and to what extent?What were the informal sources used by persons to manage the social and emotional problems of the crisis? Which areas of need were most adequately and which least adequately served by local service providers? These and similar questions must be addressed, if services are to be improved in rural areas.

We also know virtually nothing about the adjustment of displaced producers in urban settings. Although some classic studies of the adjustment of rural migrants to urban settings have been completed (e.g., Schwarzweller et al. 1971), additional analyses involving today's displaced producers are essential because of the changes that have recently occurred in the economy (see Drucker [1986] for a summary of international changes in the economy that are affecting the United States and other nations as well). Such changes have resulted in the loss of a large number of the manufacturing and similar jobs in the United States (U.S. Department of Commerce 1986) that have traditionally provided employment for rural-to-urban migrants in the United States, and as a result, it is unclear whether the most recent rural-to-urban migrants will be able to as rapidly adjust to urban settings as their predecessors.

As noted previously, most analyses of communities have concentrated on the implications of alternative factors for individuals within communities. It is essential to establish longitudinal studies of rural services and institutions in which the service and institutional organizations are the focus of the analysis. Unless such analyses are completed, it is extremely difficult to know what aspects of services and institutions should be strengthened through policy and other actions. What are the effects of the loss of a substantial number of residents on rural services and rural institutions (both formal and informal)? How extensive is the loss of leadership in institutions in rural areas due to the crisis and what effects has this loss had on the institutions involved? As for producers and rural residents in general, such questions cannot be answered without the careful collection of baseline data and the monitoring of change in organizations and institutions over time. The fact that research on community services is generally inadequate (see the recent reviews by Doeksen and Peterson [1987] and Cigler [1987]) provides an additional indication of the extent of the need for careful research that monitors the changes in rural services and institutions that occur during periods of rapid change.

In addition to the critical need for research to enhance the ability of decision makers to make appropriate and effective policies to address the needs resulting from the crisis, a number of research questions have been raised by the crisis, the answers to which are basic to the improvement of the general state of knowledge in the social sciences. Among the most important questions are those related to the impacts of rapid social change on rural communities. In many ways, the farm crisis has made agriculturally dependent communities *laboratories* for the study of social change and economic decline. In fact, recalling the quote by Clawson (see the Introduction) that social science has failed to adequately establish the social factors related to decline, it may be that the crisis provides the best opportunity in several decades to examine the determinants, concomitants, and consequences of economic and demographic decline. Although the task of chronicling the decline of rural communities is one which many rural social scientists might wish to avoid, the opportunity to advance the state of knowledge related to the interrelated events resulting from a major shift in the rural economy should not be avoided, for only by improving the state of knowledge can means be devised to assist rural areas in avoiding such decline in the future.

In addition, the fact that the major events related to the crisis are largely ones of decline may allow major contributions to be made to one of the most neglected topics in social science research—the impacts of economic decline and population loss on the socioeconomic structure of rural areas (Murdock et al. 1986). Social scientists have tended to emphasize patterns of growth rather than decline, and thus the theoretical and empirical work examining such processes of growth as modernization, urbanization, industrialization, and economic growth has been voluminous while that examining economic decline has been very limited (Hawley 1986). In many areas of social science theory as well, the lack of development of the concepts related to decline is evident. For example, in human ecology the concepts related to growth are well developed under the general rubric of ecological expansion (Hawley 1986). Expansion is assumed to involve increases in technology and organizational complexity that lead to increases in the ability of a population to further exploit its environmental resource base. The concept of decline, or ecological contraction, however, is usually described as merely the reversal of expansion (Hawley 1950), a premise that seems untenable given that technological growth rather than decline has been associated with the long-term decline in organizations and populations in rural areas. The events occurring during the current crisis thus offer a unique opportunity not only to empirically evaluate the factors involved in socioeconomic change but also to develop and test socioeconomic theories of decline as well.

A number of other less encompassing yet important questions are also addressable by analyzing the events of the farm crisis. What are the economic multipliers accompanying economic decline? What are the secondary effects related to farm failure? Are the effects of economic decline linear or do they show relationships that vary by the extent and cause of the decline? The answers to such questions have implications both for economic theory and for the mode of analysis used to assess development theories.

What are the implications of population loss for the future of demographic change and social relations in rural areas? Although the loss of young adult migrants (see chapter 7) clearly increases the age of the population and reduces the base for future fertility, rates of social interaction, the socialization of children, and other factors may be affected by the demographic context in which such events occur, but analysis of such factors at various levels and rates of population decline has not been completed.

What is the volume of business at which rural communities in different contexts are likely to experience exponential declines in the number of businesses and in the mix of businesses? Although literature delineating the likely structure of businesses in places of different population sizes is widely available, the patterns of decline associated with different rates of decline are less well known. What are the minimum levels of demand at which different types of public services become cost ineffective and at which levels is consolidation most and least advantageous? Again, although an extensive literature on these topics is available (e.g., Doeksen and Peterson 1987; Reid et al. 1984), it remains unclear at what levels economies of scale begin and end and when consolidation of services is more or less efficient than the use of smaller, geographically separated facilities.

What are the long-term social and psychological effects of failure in a rural context? The loss of jobs is not unique to rural settings, but the labelling accompanying such loss may be more extensive in a rural setting. On the other hand, persons in rural settings may also show greater concern for the recovery and restoration of individual social and self esteem. Research to establish the relative magnitude of such effects would extend our base of knowledge related to the costs and benefits of small intimate interaction groups. These and numerous other questions can be addressed by examining the causes, concomitants, and consequences of the farm crisis. It is evident that the need for research both to guide policy formation and to extend our knowledge base is extensive and should occupy a central position in the social science research agenda in the coming years.

Conclusions

In this volume we have attempted to describe the farm financial crisis in the United States, to discern its magnitude, to describe the distribution and incidence of financial difficulty among different types of producer groups, to delineate its individual and community impacts, and to estimate its long-term effects on agriculturally dependent rural areas. We have also attempted to outline needed areas for policy formulation and implementation and the areas and potential for research related to the crisis. What the final effects of the crisis will be cannot yet be determined, but the impacts are likely to leave rural America in a weakened condition. The challenge for rural-oriented social scientists, policy makers, and most importantly, the residents of rural areas in the coming years is to ensure that the base of information, the forms of policies and resources required, and the level of commitment to the survival of rural areas and the welfare of rural residents are kept at the forefront of research, policy, and social agendas for both rural areas and for the nation as a whole.

Technical Appendix

In several places is this volume, data are reported from the surveys of farmers and ranchers, business operators, and other rural residents in North Dakota and Texas conducted by the editors from 1985 through 1987. In this brief appendix, we present a summary of the survey methodologies used in the collection of these data. Readers who require additional information on the survey methodologies used or who wish to obtain copies of the survey instruments may contact either editor.

In March through May of 1985, random sample telephone surveys of producers in North Dakota and Texas were conducted. The sampling frames for these surveys were taken from those used by governmental agencies involved in periodic surveys of producers in the two states. Common survey instruments were used in the two states. A total of 1,953 operators were interviewed (933 in North Dakota and 1,020 in Texas); the overall response rate was 75 percent (70 percent in Texas and 77 percent in North Dakota). Respondents were screened to obtain a sample of persons having farming or ranching as their major economic enterprise and for whom the financial situation in agriculture was likely to have long-term consequences. Thus, interviews were limited to persons under 65 years of age who operated a farm or ranch at the time of the interview, who had gross farm sales of more than $2,500 in 1984, and who considered farming to be their primary occupation. Standard telephone survey techniques involving three callbacks at different times of the day and different days of the week were used. The sample sizes were sufficient to allow response patterns to be estimated within 10 percent of the likely population response with a 95 percent level of confidence. Comparisons of the characteristics of the survey respondents to those for the farm population in 1980 and the characteristics of farm operators from the 1982 Census of Agriculture indicated that the respondents were generally representative of producers operating commercial-sized farms in the two states (Murdock et al. 1985; Leholm et al. 1985).

The questionnaire elicited information on a wide variety of operator and financial characteristics, including the year the operator began farming, length of residence in the area, employment experience outside the area and outside of farming, size of operation (both in terms of acres and sales), type of ownership, land-tenure patterns, market-use patterns, use of farm financial techniques and of Extension Service and Experiment Station training and publications, involvement in the community, and community organizational leadership. Data were also collected on farm financial characteristics, including levels of personal and farm debts and assets (current, intermediate-term, and long-term), sources of loans, the extent to which producers were current on loan payments, net and gross farm income (as reported on income tax form 1040F), off-farm income (by source), off-farm employment experience, levels of satisfaction with farming, and levels of concern with present financial conditions in farming. Demographic characteristics included age, sex, family size, education, race/ethnicity, and marital status. The instruments also obtained information on relatives or others who would always know the producer's residence even if he/she no longer lived at the residence occupied at the time of the survey. This latter item was collected to allow for subsequent follow-up studies. The interview took approximately 30 minutes to complete.

In the spring and summer of 1986, follow-up surveys of producers in each of the states were completed. These surveys were of the same basic form as those used in 1985, but in addition to the information collected in 1985, these instruments solicited information on producers' evaluations of the impacts of the crisis on their communities, and on themselves and their families. As with the original survey, a telephone interview of approximately 30 minutes was again used with a minimum of three callbacks. In addition to standard telephone interview procedures, a letter was sent prior to the survey to thank producers for their responses in the preceding year, to offer them copies of reports resulting from the preceding year's study, and to ask them to have their financial records available to limit the time needed for the survey.

In Texas, 961 (94 percent) of the 1,020 producers who were interviewed in 1985 were recontacted in 1986. Of the 961 contacted, a total of 815 interviews were completed (85 percent of the 961 and 79.9 percent of the original sample). The completed interviews included 791 farmers who were still farming and 24 persons who had left farming during the year. An additional 137 respondents refused to be interviewed, and 9 producers had either died between 1985 and 1986 or were medically unable to participate in the survey. Of the 137, it was possible to establish that 116 were still farming and 21 had left agriculture.

In North Dakota, 759 (81.4 percent) of the 933 operators interviewed in 1985 were reinterviewed in 1986 with the remaining 174 consisting of persons who refused (99 respondents), could not be contacted (53 respondents), had ceased to operate farms (18 respondents), or were deceased (4 respondents). Because the total number of persons who had left farming who agreed to be reinterviewed was relatively small—although in both North Dakota and Texas it could be determined that the total proportion of producers leaving farming (both those who would and those who would not agree to be reinterviewed) was between 4 and 5 percent, a substantial percentage for a single year—data from the reinterview are used primarily to address specific questions regarding producers' responses in general. Responses for former producers, are derived from the North Dakota survey of displaced producers described below.

The former farmer survey was conducted during the latter part of 1986 in North Dakota only. A list of 432 producers who had quit farming was selected from lists obtained from agencies that have frequent contacts with producers. Because neither a uniform nor comprehensive list of former producers could be obtained, this was not a random sample and may be biased in unknown ways. Given the sparcity of data on former producers, however, this represents a unique data set, which we believe is useful for discerning the likely characteristics and responses of former producers.

Of the 432 producers for whom names were obtained, 260 were contacted by phone and the remainder were mailed questionnaires. Altogether, 169 useable questionnaires (39.1 percent of the total 432) were obtained. Although this response rate is much lower than in the other surveys and much lower than desired, these former producers showed little desire to be interviewed. For many, the experience of leaving farming was extremely painful and they did not wish to complete a survey about the events related to it, despite repeated attempts to obtain a response.

The questionnaire for this survey asked questions related to the characteristics of their farm during its last years of operation, about steps they had taken to restructure their farming operations, and about the general impacts of the crisis on them and their communities. Although limited in terms of its generalizability, the data from this survey provide information not available from other sources. This survey took approximately 20 minutes to complete.

The final two surveys used in the analyses in this volume involved surveys of business operators, former business operators, and community residents in 9 communities (6 in North Dakota and 3 in Texas) conducted in the summer and fall of 1986 in North Dakota and Texas. These communities ranged in population size from 1,700 to 16,000. In all communities an attempt was made to interview all current business operators, to locate and interview former business operators, and to interview approximately 100 residents in each community. For community residents, this number of respondents was of a sufficient size to provide 95 percent confidence that the estimate was within 10 percent of the population response in each community. Resident interviews were restricted to persons who were not business operators and who did not operate a farm. All three surveys were restricted to persons who were less than 65 years of age and not retired and could thus be assumed to depend on the community for a majority of their livlihood. The business operators' surveys were administered using a dropoff-and pickup-technique; operators were interviewed personally to obtain missing or incomplete information. Surveys of former business operators were also administered with a dropoff- and pickup-technique for those remaining in the communities and by mail and/or phone for those who had relocated. Telephone interviews were used to obtain information from community residents.

In the completion of the business operator survey, 1,417 businesses were contacted in the 9 communities. Of these operators, 715 or 50.5 percent responded with response rates varying from 39.6 percent to 63.4 percent among the communities. This relatively low response rate was the result of the detailed financial information requested in the survey instrument. Many operators simply refused to provide such detailed information.

A total of 143 former business operators were identified, and 77 (53.9 percent) responded. As with former producers, many former business operators did not wish to respond to a survey which required them to relive a painful experience. A total of 1,153 residents were contacted in the 9 communities with 829 or 71.9 percent responding to the survey. Response rates varied from 59.8 percent to 87.0 percent among the communities.

The current and former business operators' questionnaires requested participants to provide information on financial and other characteristics of their business. Questions were included on the number and types of employees, customers' areas of residence, inventory acquisition, and similar items. In addition, the questionnaire asked the respondents for their perceptions of the major reasons for the farm crisis, the impacts of the crisis on their own and other businesses in the community, and the impacts of the crisis on their personal lives. The survey of current business operators also asked respondents to evaluate services in their community, to list the effects of the crisis on community services, and to describe what—if any—actions they had taken in their businesses in response to the crisis. The survey of residents examined items similar to those in the business surveys with the exception of the detailed business-related questions. All three surveys obtained detailed information on the demographic and economic characteristics of the respondent and the respondent's household. Respondents were able to complete the current business operators' surveys in about 40 minutes, the residents' survey in 30 minutes, and the former business operators' survey in about 20 minutes.

Although these surveys are limited in several regards, we believe that, taken together, they provide relatively unique data related to the farm crisis. We know of no other more comprehensive set of data that includes information on both community residents' and producers' responses to the crisis collected in two states using common research instruments. Despite the utility of these data, it is evident that data with wider geographical generalizability are essential to adequately address the determinants, concomitants, and consequences of the farm financial crisis.

References

Albrecht, D.E., and Ladewig, H., "Corporate agriculture and the family farm." *The Rural Sociologist* 2(6): 376–383, 1982.

Albrecht, D.E. and Murdock, S.H., "Toward a human ecological perspective on part-time farming." *Rural Sociology* 49(3): 389–411, 1984.

Albrecht, D.E., and Murdock, S.H., "In defense of ecological analyses of agricultural phenomena: a reply to Swanson and Busch." *Rural Sociology* 50(3): 437–456, 1985a.

Albrecht, D.E., and Murdock, S.H. *The Consequences of Irrigation Development in the Great Plains*. Department of Rural Sociology Technical Report 85–1. College Station: Texas Agricultural Experiment Station, 1985b.

Albrecht, D.E., and Murdock, S.H. *The Sociology of U.S. Agriculture: An Ecological Perspective*. Ames: Iowa State University Press (in press).

Albrecht, D.E., Murdock, S.H., Hamm, R.R., and Schiflett, K.L. *The Farm Crisis in Texas: Changes in the Financial Condition of Texas Farmers and Ranchers, 1985–86*. TAES Technical Report No. 87–3. College Station: Texas Agricultural Experiment Station, 1987a.

Albrecht, D.E., Murdock, S.H., Hamm, R.R., and Schiflett, K.L. *Farm Crisis: Impact on Producers and Rural Communities in Texas*. TAES Technical Report No. 87–5. College Station: Texas Agricultural Experiment Station, 1987b.

Avery, D. *World Food Productivity: Rising Fast*. Report 967-AR. Washington, D.C.: U.S. Department of State, Bureau of Intelligence and Research, 1984.

Baker, C.B. *Current Financial Stress: Sources and Structural Implications for U.S. Agriculture*. A.E. Res. 87–1. Ithaca, N.Y.: Cornell University, Department of Agricultural Economics, 1987.

Barnes, P., and Casalino, L. *Who Owns the Land? A Primer on Land Reform in the U.S.A.* Berkeley, California: Center for Rural Studies, 1972.

Barrows, R., Jesse, E., Jones, B., Klemme, R., Pulver, G., and Saupe, W. *Financial Status of Wisconsin Farming, 1986*. Madison, Wisconsin: University of Wisconsin, Department of Agricultural Economics, 1986.

Barry, P.J. *Financial Stress in Agriculture: Policy and Financial Consequences*. Department of Agriculture, AE4621. Champaign: University of Illinois at Urbana, 1986.

Bender, L.D., Green, B.L., Hady, T.F., Kuehn, J.A., Nelson, M.K., Perkinson, L.B., and Ross, P.J. *The Diverse Social and Economic Structure of Nonmetropolitan America*. Economic Research Service, Rural Development Research Report Number 40. Washington, D.C.: United States Department of Agriculture, 1985.

Berardi, G.M., and Geisler, C.C. *The Social Consequences and Challenges of New Agricultural Technologies*. Boulder, Colorado: Westview Press, 1984.

Bertrand, A.L. *Rural Sociology*. New York: McGraw-Hill Book Company, 1958.

Bertrand, A.L., "Rural social organizational implications of technology and industry." Chapter 5 in T.R. Ford (ed.), *Rural U.S.A.: Persistence and Change*. Ames: Iowa State University Press, 1978.

Boehlje, M., and Eidman, V., "Financial stress in agriculture: implications for producers." *American Journal of Agricultural Economics* 65: 937–944, 1983.

Brake, J. R., "Financial crisis in agriculture: discussion." *American Journal of Agriculture Economics* 65(5): 953–954, 1983.

Brake, J.R., and Boehlje, M.D., "Solutions (or resolutions) of financial stress problems from the private and public sectors." *American Journal of Agricultural Economics* 67(5): 1123–1128, 1985.

Brake, J.R., and Boehlje, M.D., "Short term transition policies to ease the financial crisis." In *The Farm Credit Crisis: Policy Options and Consequences*. Washington, D.C.: USDA, Extension Service and the Farm Foundation, 1986.

Breimyer, H.F., "The changing American farm." *The Annals of the American Academy of Political and Social Science* 429: 12–22, 1977.

Brewster, J.M., "The machine process in agriculture and industry." *Journal of Farm Economics* 32(1): 69–81, 1950.

Brown, D.L., and Beale, C.L., "Diversity in post-1970 population trends." Pp. 27–71 in A.H. Hawley and S.M. Mazie (eds.), *Nonmetropolitan America in Transition*. Chapel Hill: The University of North Carolina Press, 1981.

Brunner, E. deS., and Kolb, J.H. *Rural Social Trends*. New York: McGraw-Hill Book Co., 1933.

Bultena, G., Lasley, P., and Geller, J., "The farm crisis: patterns and impacts of financial distress among Iowa farm families." *Rural Sociology* 51(4): 436–448, 1986.

Busch, L., and Lacy, W.B. *Science, Agriculture, and the Politics of Research*. Boulder, Colorado: Westview Press, 1983.

Canadian Department of Employment and Immigration. "Canadian Rural Transition Program." Informational Brochure, Ottawa, Canada: Canadian Department of Employment and Immigration, 1986.

Carlin, T.A., and Ghelfi, L.M., "Off-farm employment and the farm sector." Pp. 270–73 in *Structure Issues of American Agriculture*. Agricultural Economics Report 438. Washington, D.C.: United States Department of Agriculture, 1979.

Christenson, J.A., and Robinson, J.W. (eds.), *Community Development in America*. Ames, Iowa: Iowa State University Press, 1980.

Cigler, B.A. *Setting Smalltown Research Priorities*. Agriculture and Rural Economics Division, Economic Research Service, U.S. Department of Agriculture, Staff Report No. AGES860818. Washington, D.C.: U.S. Government Printing Office, 1987.

Clawson, M., "The dying community: the natural resource base," Pp. 55–83 in A. Gallaher, Jr., and H. Padfield, (eds.), *The Dying Community*. Albuquerque, N.M.: University of New Mexico Press, 1980.

Clifford, W.B., Miller, M.K., and Stokes, C.S., "Rural urban differences in mortality in the United States, 1970 to 1980." In D. Jahr, J.W. Johnson, and R.C. Wimberly (eds.), *New Dimensions in Rural Policy: Building Upon Our Heritage.* Washington, D.C.: U.S. Government Printing Office, 1986.

Cochrane, W.W. *The Development of American Agriculture: A Historical Analysis.* Minneapolis, M.N.: University of Minnesota Press, 1979.

Coughenour, C.M., and Swanson, L., "Work statuses and occupations of men and women in farm families and the structure of farms." *Rural Sociology* 48(1): 23–43, 1983.

Deavers, K.L., Hoppe, R.A., and Ross, P.J., "Public policy and rural poverty: a view from the 1980's." *Policy Studies Journal* 15(2): 291–309, 1986.

Dixon, W.J., Brown, M.B., Engelman, L., Frane, J.W., Hill, M.A., Jennrich, R.F., and Toporek, J.D. *BPMD Statistical Software, 1981.* Berkeley: University of California Press, 1981.

Dobson, W.D., Barnard, F.L., and Graves, B. *Results of Indiana Farm Finance Survey.* West Lafayette, Indiana: Purdue University, Department of Agricultural Economics, 1985.

Doeksen, G.A., "The agricultural crisis as it affects rural communities." *Journal of the Community Development Society* 18(1): 78–88, 1987.

Doeksen, G.A., and Peterson, J. *Critical Issues in the Delivery of Local Government Services in Rural America.* Agriculture and Rural Economics Division, Economic Research Service, U.S. Department of Agriculture, ERS Staff Report AGES860917. Washington: U.S. Government Printing Office, 1987.

Dorner, P., "Technology and U.S. agriculture." Pp. 73–86 in G.F. Summers (ed.), *Technology and Social Change in Rural Areas: A Festschrift for Eugene A. Wilkening.* Boulder, Colorado: Westview Press, 1983.

Doye, D.G., Jolly, R.W., and Choat, D. *Agricultural Restructuring Requirements by Farm Credit System District.* Staff Report 87-SR34. Ames, Iowa: Iowa State University, The Center for Agricultural and Rural Development, 1987.

Drabenstott, M., Henry, M., and Gibson, L., "The rural economic policy choice." *Economic Review* (January) 72: 41–58, 1987.

Drucker, P.F., "The changed world economy." *Foreign Affairs* Spring: 768–791, 1986.

Duncan, M., and Harrington, D.H., "Farm financial stress: extent and causes" In *The Farm Credit Crisis: Policy Options and Consequences.* Washington, D.C.: USDA, Extension Service and Farm Foundation, 1986.

Duncan, O.D., and Reiss A.J. *Social Characteristics of Urban and Rural Communities.* New York: J. Wiley and Sons, 1956.

Dunn, J.C., "Rural Farm Population Loss: Economic and Demographic Implications for North Dakota's State Planning, Region 6." Unpublished M.S. thesis, Department of Agricultural Economics, North Dakota State University, Fargo, North Dakota, 1987.

Ekstrom, B.L., and Leistritz, F.L. *Rural Community Decline and Revitalization: An Annotated Bibliography.* New York: Garland Publishing, 1988

Ekstrom, B.L., Leistritz, F.L., Vreugdenhil, H.G., and Leholm A.G., "Farm household's adjustments to changing economic conditions: highlights of 1986 farm survey." *North Dakota Farm Research* 44(3): 17–21, 27, 1986.

Engels, R.A., "The metropolitan/nonmetropolitan population at mid-decade." Paper presented at the annual meetings of the Population Association of America, San Francisco, 1986.

Extension Service. *The Farm Credit Crisis: Policy Options and Consequences.* Washington, D.C.: USDA, Extension Service, 1986.

Federal Deposit Insurance Corporation. *Annual Report.* Washington, D.C.: Federal Deposit Insurance Corporation, 1987.

Flaim, P.O., and Sehgal, E. *Displaced Workers, 1979–83.* Bulletin 2240. Washington, D.C.: U.S. Department of Labor Statistics, 1985.

Fiske, J.R., Batte, M.T., and Richenbacker, S.L. *Factors Influencing Currentness of Debt Payments for Ohio Commercial Farmers.* ESO No. 1291. Columbus, Ohio: Ohio State University, Department of Agricultural Economics and Rural Sociology, 1986.

Gee, W. *The Social Economics of Agriculture.* New York: Macmillan Co., 1942.

Gillete, J.M. *Rural Sociology.* New York: Macmillan Co., 1936.

Ginder, R.G. *The Structure of Production Agriculture and the Farm Debt Crisis.* Ames, Iowa: Iowa State University, Economics Department, 1985.

Goreham, G.A., Leistritz, F.L., and Rathge, R.W. *Trade and Marketing Patterns of North Dakota Farm and Ranch Operators.* Agricultural Economics Miscellaneous Report No. 98. Fargo: North Dakota State University, 1986.

Goreham, G.A., Leistritz, F.L., Rathge, R., Ekstrom, B., "Implications of trade and market patterns of North Dakota farm and ranch operators." *North Dakota Farm Research* 44(4): 23–27, 1987.

Graham, K.H., "A Description of the Transition Experiences of 28 New York State Farm Families Forced From Their Farms: 1982–1985." Unpublished M.S. thesis. Ithaca, New York: Cornell University, Department of Agricultural Economics, 1986.

Green, G.P., and Heffernan, W.D., "Economic dualism in American agriculture." *Southern Rural Sociology* 2: 1–10, 1984.

Guither, H.D., "Factors influencing farm operator's decisions to leave farming." *Journal of Farm Economics* 45(3): 567–576, 1963.

Guither, H.D., Jones, B.J., Martin, M.A., and Spitze, G.F. *U.S. Farmers' Views on Agricultural and Food Policy.* North Central Regional Extension Publication 227 (Res.Pub. 300). Urbana: University of Illinois, 1984.

Hardesty, S.D. *1986 Michigan Farm Finance Survey: Final Report.* Staff Paper No. 86–33. Lansing, Michigan: Michigan State University, Department of Agricultural Economics, 1986.

Harl, N.E., "The people and the institutions: an economic assessment." Pp. 71–89 in *Increasing Understanding of Public Problems and Policies—1986.* Oak Brook, Illinois: Farm Foundation, 1986.

Harrington, D., and Carlin, T.A. *The U.S. Farm Sector: How is it Weathering the 1980's?* Agricultural Information Bulletin No. 506. Washington, D.C.: USDA Economic Research Service, 1987.

Harrington, D., and Stam, J.M. *The Current Financial Condition of Farmers and Lenders.* Agricultural Information Bulletin No. 490. Washington, D.C.: USDA Economic Research Service, 1985.

Harris, M., "Landless farm people in the United States." *Rural Sociology* 6(2): 107–116, 1941.

Hawley, A.H. *Human Ecology: A Theory of Community Structure.* New York: Ronald Press, 1950.

Hawley, A.H. *Human Ecology: A Theoretical Essay.* Chicago: University of Chicago Press, 1986.

Heady, E.O. *Agricultural Policy Under Economic Development.* Ames, Iowa: Iowa State University Press, 1962.

Heffernan, J.B., and Heffernan, W.D., "The effects of the agricultural crisis on the health and lives of farm families." Columbia, Mo.: University of Missouri, Department of Rural Sociology, 1985a.

Heffernan, W.D., and Heffernan, J.B., "Testimony prepared for a hearing of the Joint Economic Committee of the Congress of the United States (September 17, 1985)." Columbia, Mo.: University of Missouri, Department of Rural Sociology, 1985b.

Heffernan, W.D., Green, G.P., Lasley, R.P., and Nolan, M.F., "Part-time farming and the rural community." *Rural Sociology* 46(2): 245–262, 1981.

Henderson, D.R., and Frank, S.D. *Farm Transition Under Financial Stress: An Ohio Case Study.* ESO 1252. Columbus, Ohio: Ohio State University, Department of Agricultural Economics and Rural Sociology, 1986.

Henry, M. Drabenstott, M., and Gibson, L., "A changing rural America." *Economic Review* (July-August) 71: 23–41, 1986.

Hightower, J., "Corporate power in rural America." Paper presented at the New Democratic Coalition Hearing, Washington, D.C., October 12, 1971.

Hill, L.D., "Characteristics of the farmers leaving agriculture in an Iowa county." *Journal of Farm Economics* 44(2): 419–426, 1962.

Hoffsommer, H., "The disadvantaged farm family in Alabama." *Rural Sociology* 2(4): 382–392, 1937.

Honadle, B.W. *Public Administration in Rural Areas and Small Jurisdictions: A Guide to the Literature.* New York: Garland Publishing, Inc., 1983.

Hottel, B., and Harrington, D.H., "Tenure and equity influences on the incomes of farmers." Pp. 97–107 in *Structure Issues of American Agriculture.* Economics, Statistics and Cooperatives Service, Agricultural Economic Report 438. Washington, D.C.: U.S. Department of Agriculture, 1979.

Johnson, C.S., Embree, E.R., and Alexander, W.W. *The Collapse of Cotton Tenancy.* Chapel Hill: University of North Carolina Press, 1935.

Johnson, J.D., Morehart, M.J., and Erickson, K., "Financial conditions of the farm sector and farm operators." *Agricultural Finance Review* 47: 1–18, 1987.

Johnson, J., Baum, K., and Prescott, R. *Financial Characteristics of U.S. Farms, 1985.* Agricultural Information Bulletin No. 495. Washington, D.C.: U.S. Department of Agriculture, Economic Research Service, 1985.

Johnson, J., Prescott, R., Banker, D., and Morehart, M. *Financial Characteristics of U.S. Farms, January 1, 1986.* Agricultural Information Bulletin No. 500. Washington, D.C.: USDA, Economic Research Service, 1986.

Jolly, R.W., and Barkema, A.D. *1985 Iowa Farm Finance Survey: Current Conditions and Changes Since 1984.* ASSIST-8. Ames, Iowa: Iowa State University, Cooperative Extension Service, 1985.

Jolly, R.W., and Olsen, D.R. *Summary of the 1986 Iowa Farm Finance Survey.* ASSIST-12. Ames, Iowa: Iowa State University, Cooperative Extension Service, 1986.

Jolly, R.W., Paulsen, A., Johnson, J.D., Baum, K.H., and Prescott R., "Incidence, intensity, and duration of financial stress among farm firms." *American Journal of Agricultural Economics* 67(5): 1108–1115, 1985.

Joseph, A., and Reinsel, R.D. *The Financial Conditions of Agriculture: An Income Analysis.* Staff Report No. AGES 860710. Washington, D.C.: USDA, Economic Research Service, 1986.

Kalbacher, J.Z., and DeAre, D., "Farm population of the United States, 1985." *Current Population Reports* P-27 No. 59, Washington, D.C.: U.S. Department of Commerce, Bureau of the Census, 1986.

Kloppenburg, J.R., Jr., and Geisler, C.C., "The agricultural ladder: agrarian ideology and the changing structure of U.S. agriculture." *Journal of Rural Studies* 1(1): 59–72, 1985.

Knutson, R.D., Penn, J.B., and Boehm, W.T. *Agricultural and Food Policy.* Englewood Cliffs, N.J.: Prentice-Hall, 1983.

Kolb, J.H., and Brunner, E., deS., *A Study of Rural Society: Its Organization and Changes.* Boston: Houghton Mifflin Co., 1935.

Korsching, P.F., and Gildner, J., (eds.), *Interdependencies of Agriculture and Rural Communities in the Twenty-first Century: The North Central Region Conference Proceedings.* Ames, Iowa: The North Central Regional Center for Rural Development, Iowa State University, 1986.

Larson, O.F., "Agriculture and the community." Pp. 147–193 in A.H. Hawley and S.M. Mazie (eds.), *Nonmetropolitan America in Transition.* Chapel Hill: University of North Carolina Press, 1981.

Lasley, P. *The Iowa Farm Poll–1985 Summary.* Ames, Iowa: Iowa State University, Department of Rural Sociology, 1985.

Lasley, P. *The Iowa Farm Poll–1987 Summary.* Ames, Iowa: Iowa State University, Department of Rural Sociology, 1987.

Lee, J. *Farm Sector Financial Problems: Another Perspective.* Agricultural Information Bulletin No. 499. Washington, D.C.: USDA, Economic Research Service, 1986.

Lee, J.E., Jr., "Some economic, social, and political consequences of the new reality in American agriculture." Pp. 87–116 in *Research in Domestic and International Agribusiness Management.* Greenwich, Conn.: JAI Press, 1982.

Leholm, A.G., Leistritz, F.L., Ekstrom, B., and Vreugdenhil, H.G. *Selected Financial and Other Socioeconomic Characteristics of North Dakota Farm and Ranch Operators.* Ag. Eco. Report No. 199. Fargo: North Dakota State University, 1985.

Leistritz, F.L., and Ekstrom, B.L. *Interdependencies of Agriculture and Rural Communities: An Annotated Bibliography.* New York: Garland Publishing, 1986.

Leistritz, F.L., Albrecht, D.E., Leholm, A.G., and Murdock, S.H., "Impact of agricultural development on socioeconomic change in rural areas." Pp. 109–138 in P.F. Korsching and J. Gildner (eds.), *Interdependencies of Agriculture and Rural Communities in the 21st Century: The North Central Region.* Ames, Iowa: The North Central Regional Center for Rural Development, 1986a.

Leistritz, F.L., Ekstrom, B.L., and Vreugdenhil, H.G. *Causes and Consequences of Economic Stress in Agriculture: Contrasting the Views of Rural Residents.* Ag. Eco. Report No. 219. Fargo: North Dakota State University, 1987b.

Leistritz, F.L., Ekstrom, B.L., Leholm, A.G., and Wanzek, J. *Families Displaced From Farming in North Dakota: Characteristics and Adjustment Experiences.* Ag. Eco. Report No. 220. Fargo: North Dakota State University, 1987a.

Leistritz, F.L., Ekstrom, B.L., and Vreugdenhil, H.G. *Selected Characteristics of Business Operators in North Dakota Agricultural Trade Centers.* Ag. Eco. Report No. 217. Fargo: North Dakota State University, 1987c.

Leistritz, F.L., Ekstrom, B.L., Vreugdenhil, H.G., and Leholm, A.G., "North Dakota farmers' views on financial assistance policies," *North Dakota Farm Research* 44(1): 31–36, 1986b.

Leistritz, F.L., Ekstrom, B.L., Wanzek, J., and Vreugdenhil, H.G., *Selected Socioeconomic Characteristics of North Dakota Community Residents.* Ag. Eco. Report No. 218. Fargo: North Dakota State University, 1987d.

Leistritz, F.L., Hardie, W.C., Ekstrom, B.L., Leholm, A.G., and Vreugdenhil, H.G. *Financial, Managerial, and Attitudinal Characteristics of North Dakota Farm Families: Results of the 1986 Farm Survey.* Ag. Eco. Report No. 222. Fargo: North Dakota State University, 1987e.

Leistritz, F.L., Leholm, A.G., Murdock, S.H., and Hamm, R.R., "The current farm financial situation: impact on farm operators and rural communities." Paper presented at the 1986 Agricultural Outlook Conference, Session #21, Washington, D.C., December 5, 1985a.

Leistritz, F.L., Vreugdenhil, H.G., Ekstrom, B.L., and Leholm, A.G. *Off-Farm Income and Employment of North Dakota Farm Families.* Agricultural Economics Misc. Report No. 88. Fargo: North Dakota Agricultural Experiment Station, 1985b.

Lines, A.E., and Morehart, M., "Financial health of U.S. farm business: a region, type, and size analysis." Paper presented at the 1986 Annual Meeting of the American Agricultural Economics Association, Reno, Nevada, July 27–30, 1986.

Lines, A.E., and Pelly R. *1985 Ohio Farm Finance Survey Results.* ESO1214. Columbus, Ohio: The Ohio State University, Department of Agricultural Economics and Rural Sociology, 1985.

Lines, A.E., and Zulauf C.R., "Debt-to-asset ratios of Ohio farmers: a polytomous multivariate logistic regression of associated factors." *Agricultural Finance Review* 45: 92–99, 1985.

Long, R.W., Reid, J.N., and Deavers, K.L. *Rural Policy Formulation in the United States.* Staff Report No. AGES 870203. Agriculture and Rural Economics Division, Economic Research Service, U.S. Department of Agriculture. Washington, D.C.: U.S. Government Printing Office, 1987.

Lonsdale, R.E., and Seyler, H.L., (eds.) *Nonmetropolitan Industrialization.* New York: John Wiley and Sons, 1979.

McKinzie, L., Baker, T.G., and Tyner, W.E. *A Perspective on U.S. Farm Problems and Agricultural Policy.* Boulder, Colorado: Westview Press, 1987.

Madden, J.P. *Economies of Size in Farming.* Agricultural Economic Report No. 107. Washington, D.C.: U.S. Department of Agriculture, 1967.

Mandle, J.R., "Sharecropping and the plantation economy in the United States South." Pp. 120–129 in T.J. Byre (ed.) *Sharecropping and Sharecroppers*. Bristol, Great Britain: Frank Cass, 1983.

Maret, E., and Copp, J.H., "Some recent findings on the economic contributions of farm women." *The Rural Sociologist* 2: 112–115, 1982.

Marousek, G., "Farm size and rural communities: some economic relationships." *Southern Journal of Agricultural Economics* 2(2): 61–65, 1979.

Mazie, S.M., and Bluestone, H. *Assistance to Displaced Farmers*. Agricultural Information Bulletin No. 508. Washington, D.C.: U.S. Department of Agriculture, Economic Research Service, 1987.

Meekhof, R.L., "Memorandum." From Chief, Inputs and Finance Branch, National Economics Division. To John E. Lee, Jr., Economics Division Research Service, USDA, Washington, D.C., June 6, 1983.

Melichar, E., "The farm credit situation and the status of agricultural banks." Paper presented at Twin Cities Agricultural Issues Round Table, St. Paul, Minnesota, April 24, 1986.

Melichar, E., "Financial condition of agricultural banks." *Agricultural Finance Review* 47: 23–39, 1987.

Miller, J.P., and Bluestone, H., "Patterns of employment change in the non-metropolitan service sector, 1969–84." Paper presented at the Annual Meeting of the Southern Regional Science Association, Atlanta, Georgia, 1987.

Mortensen, T.L., "Characteristics and determinants of delinquency of North Dakota farm operators." Unpublished M.S. thesis. Fargo, North Dakota: North Dakota State University, Department of Agricultural Economics, 1987.

Murdock, S.H., and Leistritz, F.L. *Energy Development in the Western United States: Impact on Rural Areas*. New York: Praeger, 1979.

Murdock, S.H., Albrecht, D.E., Hamm, R.R., Leistritz, F.L., and Leholm, A.G., "The farm crisis in the great plains: implications for theory and policy development." *Rural Sociology* 51(4): 406–435, 1986a.

Murdock, S.H., Hamm, R.R., Albrecht, D.E., Thomas, J.K., and Johnson, J. *The Farm Crisis in Texas: An Examination of the Characteristics of Farmers and Ranchers Under Financial Stress in Texas*. Department of Rural Sociology Technical Report 85–2. College Station: The Texas Agricultural Experiment Station, 1985.

Murdock, S.H., Jones, L.L., Hamm, R.R., and Leistritz, F.L. *The Texas Assessment Modeling System (TAMS) User's Guide*. TAES Technical Report No. 87–1. College Station, Texas: Texas Agricultural Experiment Station, 1987a.

Murdock, S.H., Leistritz, F.L., and Hamm, R.R., "The state of socioeconomic impact analysis in the United States: limitations and opportunities for alternative futures." *Journal of Environmental Management* 23: 99–117, 1986b.

Murdock, S.H., Leistritz, F.L., Leholm, A.G., Hamm, R.R., and Albrecht, D.E., "Impacts of the farm crisis on a rural community." *Journal of the Community Development Society* 18(1): 30–49, 1987b.

National Resources Committee. *Farm Tenancy: Report of the President's Committee*. Washington, D.C.: U.S. Government Printing Office, 1937.

Navarro, V., "The political and economical determinants of health and health care in rural America." *Inquiry* 13: 111–121, 1976.

Neal, E.E., and Jones, L.W., "The place of the Negro farmer in the changing economy of the cotton south." *Rural Sociology* 15(1): 30–41, 1950.

Office of Technology Assessment. *Technology, Public Policy and the Changing Structure of American Agriculture.* Washington D.C.: Office of Technology Assessment, 1986.

Otto, D. *Analysis of Farmers Leaving Agriculture for Financial Reasons: Summary of Survey Results from 1984.* Ames, Iowa: Iowa State University, Cooperative Extension Service, 1985.

Paarlberg, D. *Farm and Food Policy: Issues of the 1980s.* Lincoln, Nebraska: University of Nebraska Press, 1980.

Pederson, G., Boehlje, M., Doye, D., and Jolly, R., "Resolving financial stress in agriculture by altering loan terms: impacts on farmers and lenders." *Agricultural Finance Review* 47: 123–137, 1987.

Petrulis, M., Green, B.L., Hines, F., Nolan, R., and Sommer, J. *How is Farm Financial Stress Affecting Rural America?* Agricultural Economics Report No. 568. Washington, D.C.: USDA, Economic Research Service, 1987.

Pfeffer, M.J., "Social origins of three systems of farm production in the United States." *Rural Sociology* 48(4): 540–562, 1983.

Popovich, M.G. *State Emergency Farm Finance, Volume 2.* Washington, D.C.: Council of State Policy and Planning Agencies, 1987.

Raup, P.M., "The crisis in agriculture." Staff paper, P85–34. St. Paul, Minn.: University of Minnesota, Department of Agricultural Economics, 1985.

Reid, J.N., Stinson, T.F., Sullivan, P.J., Perkinson, L.B., Clarke, M.P., and White-head, E. *Availability of Selected Public Facilities in Rural Communities: Preliminary Estimates.* Economic Development Division, Economic Research Service, U.S. Department of Agriculture. Washington, D.C.: U.S. Government Printing Office, 1984.

Reimund, D., "Form of business organization." In ESCS *Structure Issues of American Agriculture.* Agricultural Economics Report 438. Washington, D.C.: U.S. Department of Agriculture, 1979.

Richardson, J.W., and Bailey D.V. *Debt Servicing Capacity of Producers in the Blacklands.* Report prepared for the Farm Credit Banks of Texas and the Texas Agricultural Experiment Station. College Station, Texas: Texas Agricultural Experiment Station, 1982a.

Richardson, J.W., and Bailey, D.V. *Debt Servicing Capacity of Producers in the Rolling Plains.* Report prepared for the Farm Credit Banks of Texas and the Texas Agricultural Experiment Station. College Station, Texas: Texas Agricultural Experiment Station, 1982b.

Richter, K., "Nonmetropolitan growth in the late 1970s: the end of the turnaround?" *Demography* 22: 245–263, 1985.

Roemer, M.I. *Rural Health Care.* St. Louis: The C.V. Mosby Company, 1976.

Rogers, E. *Diffusion of Innovations.* New York: Free Press, 1983.

Salant, P., Smale, M., and Saupe, W. *Farm Viability: Results of the USDA Family Farm Surveys.* Rural Development Research Report No. 60. Washington, D.C.: USDA, Economic Research Service, 1986.

Schlebecker, J.T. *Whereby We Thrive: A History of American Farming, 1607–1972.* Ames: Iowa State University Press, 1975.

Schmieder, E., "Will history repeat in rural America?" *Rural Sociology* 6(4): 291–299, 1941.

Schuler, E.A., "The present social status of American farm tenants." *Rural Sociology* 3(1): 20–33, 1938.

Schwarzweller, H.K., Brown, J.S., and Mangalam, J.J. *Transition.* University Park, Pennsylvania: The Pennsylvania State University Press, 1971.

Shepard, L.E., and Collins, R.A., "Why do farmers fail? farm bankruptcies 1910–78." *American Journal of Agricultural Economics* 64: 609–615, 1982.

Smith, E.G., Richardson, J.W., and Knutson, R.D. *Cost and Pecuniary Economies in Cotton Production and Marketing: A Study of Texas High Plains Cotton Producers.* B-1475. College Station, Texas: Texas Agricultural Experiment Station, 1984.

Stinson, J., Fortenberry, J., Sigalla, F., and Conlon, T. *Governing the Heartlands: Can Rural Communities Survive the Farm Crisis?* A report prepared for the Senate Subcommittee on Intergovernmental Relations. Washington, D.C.: U.S. Congress, 1986.

Stockdale, J.D., "Who will speak for agriculture?" Pp. 317–327 in D.A. Dillman and D.J. Hobbs (eds.), *Rural Society in the U.S.: Issues for the 1980s.* Boulder, Colorado: Westview Press, 1982.

Swanson, L.E., and Busch, L., "A part-time farming model reconsidered: a comment on a POET model." *Rural Sociology* 50(3): 427–436, 1985.

Todd, R.M., "Taking stock of the farm credit system: riskier for farm borrowers." *Quarterly Review (Federal Reserve Bank of Minneapolis)* Fall: 14–24. Minneapolis, Minnesota: Federal Reserve Bank, 1985.

Tweeten, L. *Foundations of Farm Policy.* Lincoln: The University of Nebraska Press, 1970.

Tweeten, L., and Brinkman, G.L. *Micropolitan Development.* Ames, Iowa: Iowa State University Press, 1976.

Uhlman, J.M., and Olson, J.K. *Planning for Rural Human Services.* Denver: Office of Human Development Services, Department of Health and Human Services, 1984.

University of Colorado. *A Study of Selected Colorado Demographic, Economic, and Social Characteristics with Reference to Agricultural and Rural Change.* Denver, Colo.: University of Colorado, College of Design and Planning, 1985.

United States Congress, Office of Technology Assessment. *Technology, Public Policy, and the Changing Structure of American Agriculture.* OTA-F-285. Washington, D.C.: United States Government Printing Office, 1986.

United States Department of Agriculture. *Agricultural Finance Outlook and Situation Report.* USDA, Economic Research Service, AFO-26. Washington, D.C.: United States Government Printing Office, 1986a.

United States Department of Agriculture. *Economic Indicators of the Farm Sector, National Financial Summary, 1985.* USDA Economic Research Service. Washington, D.C.: United States Government Printing Office, 1986b.

United States Department of Agriculture. *Farm Population Estimates, 1910–70.* Statistical Bulletin No. 523. Washington, D.C.: United States Government Printing Office, 1973.

United States Department of Agriculture. *The Current Financial Condition of Farmers and Farm Lenders.* Economic Research Service, Agricultural Information Bulletin No. 490, Washington, D.C.: U.S. Department of Agriculture, 1985.

United States Department of Commerce, Bureau of the Census. *1982 Census of Governments.* Washington, D.C.: United States Government Printing Office, 1984–1985.

United States Department of Commerce, Bureau of the Census. *County Statistics File 2 (CO-STAT 2).* Tape and Technical Documentation. Washington, D.C.: United States Department of Commerce, Bureau of the Census, Data User Services Division, 1986a.

United States Department of Commerce, Bureau of the Census. "Geographic mobility." *Current Population Report* P-20, No. 368. Washington, D.C.: United States Government Printing Office, 1981. United States Department of Commerce, Bureau of the Census. "Geographic mobility." *Current Population Report* P-20, No. 377. Washington, D.C.: United States Government Printing Office, 1983.

United States Department of Commerce, Bureau of the Census. *1982 Census of Agriculture: United States Summary and State Data.* Washington, D.C.: United States Government Printing Office, 1984a.

United States Department of Commerce, Bureau of the Census. "Geographic mobility." *Current Population Report* P-20, No. 384. Washington, D.C.: United States Government Printing Office, 1984b.

United States Department of Commerce, Bureau of the Census. "Geographic mobility." *Current Population Report* P-20, No. 393. Washington, D.C.: United States Government Printing Office, 1984c.

United States Department of Commerce, Bureau of the Census."Geographic mobility." *Current Population Report* P-20, No. 407. Washington, D.C.: United States Government Printing Office, 1986b.

United States Department of Commerce, Bureau of the Census. *Historical Statistics of the United States: Colonial Times to 1970, Part 1 and Part 2.* Washington, D.C.: United States Government Printing Office, 1975.

United States Department of Commerce, Bureau of the Census. "Vol. 1, characteristics of the population: chapter A, number of inhabitants; chapter B, general population characteristics; chapter C, general social and economic characteristics; and chapter D, detailed population characteristics." *U.S. Censuses of Population and Housing for 1930, 1940, 1950, 1960, 1970, and 1980.* Washington, D.C.: United States Government Printing Office, 1930–1983.

Watt, D.L., Larson, J.A., Pederson, G.D., and Ekstrom, B.L. *The Financial Status of North Dakota Farmers and Ranchers: January 1, 1985, Survey Results.* Agricultural Economics Report No. 207. Fargo, North Dakota: North Dakota State University, Department of Agricultural Economics, 1986.

Wilkening, E., and Gaaleski, B. *Family Farming in Europe and America.* Boulder, Colorado: Westview Press (in press).

Wright, J.S., and Lick, D.W., "Health in rural America: problems and recommendations." Pp. 461–469 in D. Jahr, J.W. Johnson, and R.C. Wimberley (eds.)., *New Dimensions in Rural Policy: Building Upon Our Heritage.* Studies prepared for the Joint Economic Committee, Congress of the United States, Washington, D.C.: U.S. Government Printing Office, 1986.

Zeichner, O., "The transition from slave to free agricultural labor in the southern states." *Agricultural History* 12(1): 22–32, 1939.

Index